Community Management of Rural Water Supply

The supply of reliable and safe water is a key challenge for developing countries, particularly India. Community management has long been the declared model for rural water supply and is recognised to be critical for its implementation and success. Based on 20 detailed successful case studies from across India, this book outlines future rural water supply approaches for all lower-income countries as they start to follow India on the economic growth (and subsequent service levels) transition.

The case studies cover state-level wealth varying from US$2,600 to US$10,000 GDP per person, and a mix of gravity flow, single village and multi-village ground-water and surface water schemes. The research reported covers 17 states and surveys of 2,400 households. Together, they provide a spread of cases directly relevant to policy-makers in lower-income economies planning to upgrade the quality and sustainability of rural water supply to meet the Sustainable Development Goals, particularly in the context of economic growth.

Paul Hutchings is a lecturer in water and sanitation at the Cranfield Water Science Institute, Cranfield University, UK.

Richard Franceys is a consultant and led the Community Water Plus project as senior lecturer in water and sanitation management at Cranfield Water Science Institute, Cranfield University, UK.

Stef Smits is a senior programme officer at IRC in the Netherlands.

Snehalatha Mekala is the regional advisor for SPLASH, South Asia, and national research coordinator for the Community Water Plus project based in Hyderabad, India.

Earthscan Studies in Water Resource Management

For more information and to view forthcoming titles in this series, please visit the Routledge website: www.routledge.com/books/series/ECWRM

Community Management of Rural Water Supply

Case Studies of Success from India

Paul Hutchings, Richard Franceys, Stef Smits and Snehalatha Mekala

LONDON AND NEW YORK

First published 2017
by Routledge

2 Park Square, Milton Park, Abingdon, Oxfordshire OX14 4RN
52 Vanderbilt Avenue, New York, NY 10017

Routledge is an imprint of the Taylor & Francis Group, an informa business

First issued in paperback 2019

© 2017 Cranfield University, IRC and Snehalatha Mekala

The right of Paul Hutchings, Richard Franceys, Snehalatha Mekala and Stef Smits
to be identified as authors of this work has been asserted by them in accordance
with sections 77 and 78 of the Copyright, Designs and Patents Act 1988.

The research on which this book is based was funded by the Australian
Government through the Australian Development Awards Research Scheme of the
Department of Foreign Affairs and Trade under an award titled Community
Management of Rural Water Supply Systems in India. The views expressed in this
report are those of the project and not necessarily those of the Australian
Government. The Australian Government accepts no responsibility for any loss,
damage or injury, resulting from reliance on any of the information or views
contained in this report.

All rights reserved. No part of this book may be reprinted or reproduced or utilised
in any form or by any electronic, mechanical, or other means, now known or
hereafter invented, including photocopying and recording, or in any information
storage or retrieval system, without permission in writing from the publishers.

Trademark notice: Product or corporate names may be trademarks or registered
trademarks, and are used only for identification and explanation without intent to
infringe.

British Library Cataloguing-in-Publication Data
A catalogue record for this book is available from the British Library

Library of Congress Cataloging in Publication Data
Names: Hutchings, Paul, 1986- author. | Franceys, R., author. | Smits, Stef,
author. | Mekala, Snehalatha, author.
Title: Community management of rural water supply : case studies of success
from India / Paul Hutchings, Richard Franceys, Stef Smits and Snehalatha Mekala.
Description: London ; New York : Earthscan from Routledge, 2017. | Series:
Earthscan studies in water resource management | Includes bibliographical
references and index.
Identifiers: LCCN 2016056824 | ISBN 978-1-138-23207-5 (hbk) |
ISBN 978-1-315-31333-7 (ebk)
Subjects: LCSH: Water-supply, Rural—India—Management—Case studies. |
Municipal water supply—India—Case studies. | Villages—India.
Classification: LCC TD303.A1 H88 2017 | DDC 363.6/1068—dc23
LC record available at https://lccn.loc.gov/2016056824

ISBN: 978-1-138-23207-5 (hbk)
ISBN: 978-0-367-33516-8 (pbk)

Typeset in Bembo
by FiSH Books Ltd, Enfield

Contents

Figures and tables

Figures

Tables

Abbreviations

CapEx	capital expenditure
CapManEx	capital maintenance expenditure
CSP	community service provider
DRA	demand-responsive approach
ESE	enabling support environment
GDP	gross domestic product
GP	Gram Panchayat
GPWSCs	Gram Panchayat Water and Sanitation Committee
HDI	Human Development Index
IEC	information, education and communication
lpcd	litres per capita per day
LSG	Local self-government
MDG	Millennium development goal
MOU	Memorandum of understanding
MVS	multi-village scheme
NGO	non-governmental organisation
NRDWP	National Rural Drinking Water Programme
O&M	operations and maintenance
OpEx	operating and minor maintenance expenditures
OpExES	operating expenditure on enabling support
OpexpDS	operational expenditure of direct support
OpexpIDS	operational expenditure of indirect support
PHED	public health engineering department
PPP	purchasing power parity
PRI	Panchayat Raj Institutions
PWS	piped water supply
QIS	qualitative information systems
RRWSA	Reformed Rural Water Supply Agency
RWS	rural water supply
SDG	sustainable development goal
SLECs	Scheme-level Executive Committee
SRPPs	sector reform pilot project
SRWSA	state rural water supply agency

SVS	single village scheme
SWAp	sector wide approach
UNICEF	United Nation International Children Emergency Fund
VLOM	village-level operation and maintenance
VWSC	village water and sanitation committee
WASMO	Water and Sanitation Management Organisation
WRA	Water Regulatory Authority

Acknowledgements

The research described in this report was undertaken through a research grant awarded under the Australian Development Research Awards Scheme of the Department of Foreign Affairs and Trade. We are most grateful to the people of Australia for supporting this investigation and hope that the findings will be of direct policy relevance to future government and donor support to the extension of rural water supply in lower-income countries.

At the commencement of the research Kurian Baby (IAS, Kerala) was leading the IRC programme in India, and we are grateful for his support and his help in bringing together our National Research Steering Committee, headed by Mr Sujoy Mojumdar (then director of RWS). The committee also included Dr Manish Kumar (WSP-SA), Mr Arugugam Kalitmuthu and Satya Narayan Ghosh (Water for People), Mr R. P. Kulkarni (Karnataka RW S &S), Dr A. J. (Viju) James (independent consultant), Dr Hemant Kumar Joshi (CCDU, Rajasthan), Mr Ravi Narayan (adviser to Arghyam) and Mr Joe Medith (Gramvikas). We are grateful for the support of the Government of India in this research, and to each of the Steering Committee members, who kindly gave of their time to comment and advise on the research.

The research also held two stakeholder meetings, in Delhi in 2013 and 2016, and three dissemination workshops with 15 states sending staff to attend the field visits to successful community-managed water supply programmes and subsequent workshops, in Kerala, in Odisha and in Punjab. We gratefully acknowledge all the community service providers, householders and staff in support entities who so kindly answered all our questions, and particularly colleagues in Kerala, Odisha (Gram Vikas's Sojan K. Thomas and Joe Medith) and Punjab.

Dr Paul Hutchings was the lead author for most of the chapters in this book. Dr Richard Franceys, being principal investigator for the research programme, contributed to all chapters. Stef Smits led on the international implications chapter and Dr Snehalatha Mekala on the monitoring and regulation and gender chapters, and both contributed to all other chapters through their comments and research insights. We are grateful for the contributions of the research teams, as follows: Professor Srinivas Chary, Ms Shaili Jasthi and Ms Swapna Uddaraju (Administrative Staff College of India, Hyderabad, Telangana); Dr Rema Saraswathy, Dr Rammohan Rao, Mr Raviprakash Madhudi with M. S.

Vaidyanathan (Centre of Excellence for Change, Chennai, Tamil Nadu); Dr Urmila Brighu and Mr Rajesh Poonia (Malaviya National Institute of Technology, Jaipur, Rajasthan); Mr Prakash Dash and Mr Pramil Panda (Xavier Institute of Social Service, Ranchi, Ranchi, Jharkhand); Ms Ruchika Shiva and Mr Depinder Kapur with Stef Smits (IRC, the Netherlands); Mr Benjamin Harris and Mr Matthias Javorszky (Cranfield University, UK).

Much valued support was also given by Dr Alison Parker and Cranfield's 'Community Water and Sanitation Group Project' students:

- 2014: we thank Mei Yee Chan, Lucie Cuadrado, Fatine Ezbakhe, Baptiste Mesa, Chiaki Tamekawa for their work on the global literature review; with special thanks to Mei Yee Chan for working on the preparation of the final research report.
- 2015: we thank Jayan Amin, Justine Denis, Benjamin Harris, Nnamdi Ibenegbu, Matthias Javorszky, Emmanuelle Maillot, Selene Tripp Mercado for their work on harmonising the household survey data and the preparation of the manual on household surveys for rural water.
- 2016: we thank Thibault Guinaldo, Christoph Leitner, Justus Nyangoka, Vincent Thomas, Julia Zeilinger for their work on the finalisation of the financial data and the development of the financial flow diagrams.

For more information about the project that resulted in this volume, please contact Paul Hutchings (p.t.hutchings@cranfield.ac.uk) or Stef Smits (smits@ircwash.org). The research project website contains the full research methodology, the 20 full case study reports along with summaries of each of the case studies, along with additional research material: www.ircwash.org/projects/india-community-water-plus-project

1 Research purpose and background

In Ghataur, a village that is 30 km from Chandigarh the state capital of Punjab, Mr Daljit Singh is both a village resident and a long-term engineer in the Department for Water Supply and Sanitation. Over the past 20 years, he has been instrumental in supporting the development of a highly successful water service in his village that delivers piped water supply to every household 24 hours a day. He acts as the public health engineering representative on the village water committee, while his wife and now daughter play key roles within that institution helping to ensure that the community-managed service is managed to the highest professional standards. The situation in Ghataur can be described as one in which a charismatic leader or, in this case, family of leaders transforms a community's water service. This is a story that has become associated with the community management model: an exceptional individual, in the right place at the right time, plays a critical role successfully managing a village's water supply arrangements leading to the associated improvements in public health and livelihoods within that community.

While this book commends the sterling work of Mr Singh and his family, it is intentionally not about such exceptional characters. This book starts from the premise that not every village is lucky enough to have a Mr Singh, but that *every* village still deserves the high-quality water service that he has helped to deliver. Instead, this book is about learning what can make community management work at scale – not just in exceptional villages – but in ordinary villages across India and beyond. It presents the results of a three-year research investigation into what works in 20 successful community management programmes operating in contemporary India. It focuses on what will be described as the 'enabling support environment' for community management, relating this to various forms of 'community service provision', while also seeking to understand what a sustainable financing approach for the community management model looks like.

The research programme on which this book is based, 'Community Water Plus', was designed to gain insights into the type and amount of support (the 'plus') that has been needed for community management to be successful, as well as into the resources implications of that 'plus', across a range of technologies and conditions. Specifically, the project has focused on the following main research question: what type, extent and style of supporting organisations are apparent in sustainable community-managed water service delivery relative to varying technical modes of supply and what have been the resource implications of their support? In order to test these ideas, the research investigated 20 cases of community management in India.

India was selected as a country at the forefront of efforts to expand access to rural water services which also has a long history of community management. As will be described throughout the book, following the scaling up of the model during the sector reforms of the 1990s and 2000s, the country is now home to a variety of community management programmes across the 29 states. However, success remains uneven, with some notable success stories, but with continued evidence of failure (James, 2004, 2011). In India, and elsewhere in the world, there remains a crucial need to understand what mechanisms for support have worked, and to develop realistically costed policies and strategies for scaling-up and strengthening support to community-managed rural water supplies. This is the gap that this book intends to fill.

Outline of the book

This book has been organised into three parts that contain a number of thematically organised chapters that follow the overall research process that underpins the book. Part I, 'Community management: background, review and challenges', is made up of four chapters. Chapter 2 provides a general introduction to the challenges and future trends in the community management of rural water services. This is followed, in Chapter 3, by a systematic review of the global evidence on the 'success factors' for community management that helps illustrate the variety of practices applied around the world. Within Chapter 4, the focus is narrowed to the main empirical focus of the book – India – with a concentrated review of the history, concepts and typologies of community management within that country. That chapter ends with a discussion about the Indian operating context that considers the differences between states in terms of rural water supply access, political economy and geography. Chapter 5 presents an overview of the research methodology used to comply the 20 case studies of reportedly successful community management programmes from India, around which this book is based.

Part II presents 'Community management case studies of success from India'. In Chapters 6–9, case study summaries and analysis are presented on sets of case studies. These are organised by the political-economy classifications of the states discussed in Chapter 4, which cover the generally, poorer, 'neo-patrimonial states', more egalitarian, 'social-democratic states' and the richer but top-down 'developmental states'. The final (fourth) category of case studies is those from the 'mountainous states' with the distinct hydrogeological challenges of delivering rural water services in that context.

The book then progresses to investigate and analyse the trends across the case studies in Part III, 'Synthesis of successful community management arrangements in India'. Here, Chapter 10 tackles the institutional dimensions of community management by developing typologies of support and service delivery models found across the case studies, while Chapter 11 analyses the financing patterns identified within the research. This section then progresses with two shorter chapters, Chapters 12 and 13, assessing the trends for two key topics: the 'gender dimensions of community management' and the 'regulation and monitoring of community management', respectively. Finally, Chapter 14 offers a discussion about the intersection between local self-government and community management within India, discusses the international implications of the research for other regions where community management is commonly followed, and provides the final conclusions from the research. The appendix provides an overview of the case study sites and a selection of the analytical tables used to compare the case studies.

Part I

Community management

Background, review and challenges

2 Community management and community management plus

The background

The expansion of safe water services to an ever larger proportion of the global population has been a cause for celebration in recent years. In 2012 the WHO and UNICEF Joint Monitoring Programme (JMP) for Water Supply and Sanitation announced the achievement of Millennium Development Goal (MDG) target 7c three years ahead of schedule (WHO and UNICEF, 2012). This meant that from 1990 to 2012 the world had halved the proportion of the global population without access to an improved drinking water source,[1] expanding access to 2.6 billion more people (WHO and UNICEF, 2013). Now, as the world enters the Sustainable Development Goal (SDG) era, the global water sector is faced with an aspiration to deliver universal access to every person on the planet by 2030 (United Nations, 2015).

Yet, in 2015, there were still 663 million people around the world who lacked basic access – eight out of ten of whom live in rural areas (WHO and UNICEF, 2015). More fundamentally, there is strong evidence that a lack of service sustainability is threatening global progress (Fisher et al., 2015; Mandara et al., 2013; Schweitzer and Mihelcic, 2012; Stalker Prokopy and Thorsten, 2009). Research shows that over 30 per cent of rural water supply infrastructure in Sub-Saharan Africa (Adank et al., 2012; Sutton, 2005) and India (Government of India, 2009; Ratna Reddy et al., 2010) are either below the designed functionality or in a non-functioning state. This situation endangers the progress made in recent decades and, if it is not addressed, makes achieving the universal aspiration of the SDGs all but impossible. In light of such challenges, this books deals with the management challenges of providing rural water supply.

In order to do so, we first conceptualise water supply as a service that undergoes a continuous life-cycle that ensures uninterrupted water supply to people. We then discuss the history of community management, as the main way through which that service is managed in the context of rural water supply in low and middle-income countries. The chapter ends with a discussion about what constitutes a good rural water service, with a focus on the effectiveness, sustainability and replicability of rural water supply – and how community management impacts on these factors.

Water supply as a service

This section sets out how we understand successful rural water services. It starts from the recognition that rural water supply needs to be seen as an on-going service rather than a system to be developed (Fonseca et al., 2010). A water service consists of access to a flow of water with certain characteristics (such as quantity, quality and continuity). For such a service to be supplied, one has to consider the physical infrastructure (the system) as well as an entity to manage that system and, in addition, other entities to monitor, support and, at some level, regulate that system.

Water services delivery can be characterised by four phases over time, as outlined in Figure 2.1. Phase one is the capital investment or implementation phase in which the physical systems are built (i.e. the development of the initial or 'new' construction of the physical system). The second phase is the recurrent cost-supported service delivery phase, in which consumers receive the desired water supply, enabled by appropriate operation and minor maintenance which in turn has to be supported by a suitable level of administration. Support to this administration is also a likely part of this phase. Depending on the technology used initially, the third phase, occurring some or many years later, is described as the capital maintenance phase where major replacements and renewal of physical assets take place. This phase is often not very discretely defined, as such activities happen in on-going steps as physical assets reach the end of their working lives, typically throughout the extended service period. Finally in the last phase, the process continues when the need for a significant upgrade of the service is required to expand or enhance the service (e.g. delivering additional boreholes as a village expands geographically or the numbers of consumers develop through population growth).

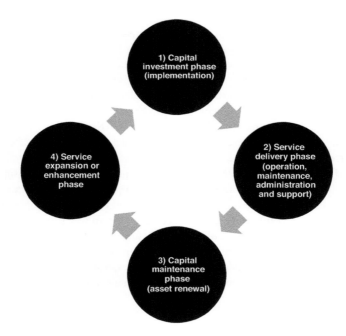

Figure 2.1 Phases in the service delivery process

Source: based on Lockwood and Smits (2011).

Community management

To provide these services, this book focuses on the future of the most common management model for rural water services – the community management approach (Van den Broek and Brown, 2015; Harvey and Reed, 2006; Moriarty et al., 2013; Schouten and Moriarty, 2003). The historical and ideological reasons behind the widespread application of this management model will be discussed throughout this chapter. However, for now, it is important to understand that community management has been the favoured approach to managing rural water services in low and lower middle income countries since at least the 1980s, playing an important role in the expansion of water services to hundreds of millions of people (Harvey and Reed, 2006; Lockwood and Smits, 2011; McCommon et al., 1990; Paul, 1987; Schouten and Moriarty, 2003). It is an approach that can be defined loosely as one in which communities should be involved in the development of water supply systems, then take ownership of them, and have overall responsibility for their on-going operation and maintenance (Harvey and Reed, 2006; Moriarty et al., 2013).

Yet there is now a growing consensus that the approach needs to be reformed (Baumann, 2006; Van den Broek and Brown, 2015). Moriarty et al. (2013, p. 329) succinctly clarify a generally held belief that community management is 'at the beginning of the end … not principally because community management has failed, but because it is reaching the limits of what can be realistically achieved in an approach based on informality and voluntarism'. Simply put, community management has conventionally been an approach in which governments and other agencies can concentrate on developing infrastructure and not worry about operating or maintaining it. This can be considered an appropriate and efficient model when the challenge is expanding access to water services but one of the primary challenges the sector now faces is to ensure the sustainable operation and maintenance of water services (Hutton and Varughese, 2016; Lockwood and Smits, 2011; Moriarty et al., 2013).

In response to this situation there have been calls to make governments (and other supporting agencies) take greater responsibility for providing continuous, on-going support to communities beyond the conventional focus on infrastructure creation (Kleemeier, 2000; Lockwood, 2002, 2004). This on-going support can take different forms and numerous terms have been used to describe it: institutional support mechanisms (Lockwood, 2002), follow-up support (Lockwood et al., 2003), post-construction support (Bakalian and Wakeman, 2009), direct support (Smits et al., 2011) and support to service providers (Smits et al., 2013). Collectively, they have become known by the term 'Community Management Plus', a phrase coined by Baumann (2006) with the 'plus' signifying the on-going support that is required to ensure community management is sustainable. Yet while the turn to community management plus has become a widely accepted shift there remains a lack of a robust guidance on how best to structure and finance such support services (Smits et al., 2011).

Community engagement in rural water supply services

Before investigating such matters we turn to the historical evolution of community management. The involvement of communities in rural water supplies has its roots in the International Decade for Drinking Water and Sanitation in the 1980s, designed to rapidly increase access to rural water supplies. Earlier approaches had favoured 'public works' or 'public health engineering', in which central government departments constructed new water supply systems in rural areas, and then tried to manage these centrally. But more often than not these systems were neglected after they were built. This approach was seen to be relatively inefficient and ineffective. In response, the 1980s gave rise to the concept of community participation – initially referring to the engagement of communities in the development of the water systems, as described in this extract from McCommon and colleagues:

> funding has been declining and many completed systems are in disrepair or have been abandoned. This state of affairs has led many experts to question whether the emphasis on centrally-managed schemes needs to be re-evaluated and a new approach taken to the provision of rural water supply as a public service. Community management has been proposed as one possible alternative strategy in view of the increasing evidence that systems are more sustainable when designed, established and operated by the community.
>
> (McCommon et al., 1990, p. 2)

However, the initial result of community participation in implementation of water systems did not yield immediate positive results, because communities were given little space to participate in key decisions, let alone take full control of the water systems. McCommon et al. again explain how these two concepts – community participation and community management – are related, but different:

> The situation did not improve markedly even when some community-based participation was encouraged, largely because community participation has been narrowly defined as the mobilization of self-help labour or the organisation of local groups to ratify decisions made by outside project planners. Externally-imposed solutions do little to build capacity, increase empowerment, or create support structures that represent the interests of users willing to maintain these rural water supply and sanitation systems on a long-term basis. Community management, as distinguished from community participation, is taken to mean that the beneficiaries of rural water services have responsibility, authority, and control over the development of such services.
>
> (McCommon et al., 1990, p. 3)

In essence, McCommon et al. (1990) identified two shifts in the conceptualisation of community-engaged water supply to community-managed water supply:

- the quality of community participation: from labour contributions and deci-sion-making on minor issues to full responsibility and decision-making on all key aspects of the services; and
- the phases in the service delivery cycle in which communities would partici-pate: from community participation in the initial development of the infrastructure only, to also their participation in the subsequent service deliv-ery and operation and maintenance of the systems.

The latter shift came from the thinking to place responsibility for operation and maintenance (O&M) onto the community, that being the first level of scale at which such activities could be undertaken. One of the first manifestations in which this thinking on community participation and community management was systematised was the village-level operation and maintenance (VLOM) approach. Under this approach communities were made fully responsible for the O&M of the water systems. It also built on what had earlier been called 'appropriate technol-ogy', and centred on basic technologies and systems that had been purposefully designed to require minimal external inputs, but primarily referred to handpumps. In practice VLOM proved to be insufficient to address the problem of sustainabil-ity as many communities were ill-prepared to take on the management responsibilities, even of the most low-cost technologies. Moreover, these – largely donor-driven – VLOM programmes favoured working directly with communities and grassroots organisations only, thereby by passing government structures. This in turn meant that after these programmes ended, communities were left alone to manage their systems.

By the end of the 1990s the discourse developed further and put emphasis on the combination of a demand-responsive approach (DRA) as well as community management. Three central elements of this combined approach were:

- A requirement that communities express a demand for services and external agencies try and respond to this. This demand was expected to be manifested through user contributions to capital costs, as that – so the argument went – would lead to a sense of ownership and hence commitment to ensuring on-going operations and use. However, Marks and Davis (2012) show that there is a threshold effect in such contribution in that it needs to be signifi-cantly high to create such a sense. In reality, it often has taken a long time for communities to mobilise such contributions and where they are made, they are often minimal (Jones, 2013). Even these minimal contributions are often waived or reduced to token contributions for the sake of speeding up infra-structure development. Real ownership, in legal terms, of assets meanwhile remained vaguely defined, if at all.
- Full cost recovery, understood to refer to user tariffs covering all operation and minor maintenance costs – which of course were only a subset of the full costs.
- Stronger and more meaningful participatory approaches. In order to achieve the previous two points, many organisations particularly non-governmental

organisations (NGOs), gradually improved the quality of participation in their projects, giving a stronger voice to communities in expressing their demand and decision-making in, for example, technology selection, tariff-setting, establishing the management model and preparing them better for their role in eventually managing the service. A range of participatory methodologies and tools were developed and specified for rural water supplies (Lammerink and De Jong, 1999; Dayal et al., 2000; Deverill et al., 2002; Bolt and Fonseca, 2001), becoming often part of the standard intervention model of these organisations. For India, the main guidelines that define the way in which communities participate are the ones set by the Ministry of Drinking Water and Sanitation (Government of India, 2013a).

Particularly, the latter element of the more meaningful participatory approaches evolved markedly in the 1990s, largely under the influence of broader developments in thinking on community participation in general. This is explained in the next section while a discussion regarding cost recovery follows that section.

Laying the foundations: from community participation to community management

The theoretical foundations for community management of rural water supplies lie in the broader work on community participation in rural development. This started from the observation that many development projects were not leading to the expected results, often because the intended user groups were not making use of certain interventions, or that these interventions had all kinds of unintended negative impacts. One of the main identified reasons for this mismatch was that many of the development projects were implemented in a top-down manner. The terminology itself implies that communities participate in another entity's programme and therefore do not have direct ownership and responsibility.

As the concept of community participation evolved, more emphasis was placed on the decision-making power of communities over development programmes. For example, Paul (1987) reports the World Bank definition of community participation: 'an active process whereby beneficiaries influence the direction and execution of development projects rather than merely receive a share of project benefits'. Appealing as such a definition of community participation may be, as a term and approach it has been used and abused by agencies, often seeing the term as window-dressing for approaches in which communities actually had very little voice. McCommon et al. (1990) list the following possible objectives of community participation in the context of development programmes:

(a) sharing project costs (beneficiaries contribute money or labour);
(b) increasing project efficiency (beneficiaries assist in project planning and implementation);
(c) increasing project effectiveness (beneficiaries have a say in project design and implementation);

(d) building beneficiary capacity (beneficiaries share in management tasks or operational responsibilities); and

(e) increasing community empowerment (beneficiaries share power and increase their political awareness and influence over developmental outcomes).

As can be seen, community participation can be understood to refer to anything from contributing money or labour to a programme to sharing in decision-making and management tasks.

The seminal work by researchers like Robert Chambers (1983) helped develop an alternative development model and methodologies which emphasised that community participation should be about providing a real voice and decision-making power to communities, so they could choose the interventions they most needed. To explain this point further, different scholars developed continuums, or ladders, of community participation, based on the original work by Arnstein (1969) in the context of social development work in the United States and further elaborated subsequently.

The challenge, as Robinson and Nolan-ITU (2002) explain, is that moving 'up' the community engagement ladder from 'tokenism' towards 'empowerment' requires 'increased capacity for information processing and learning, problem solving and resolving conflict'. The challenge to this is the obvious, often overlooked, realisation that poor communities are not only poor economically but also tend to be poor in terms of social and institutional capital. For example, they may lack democratic structures for decision-making and local elites may capture a disproportional amount of benefits from water projects. Again, the majority of the

Table 2.1 Ladders of community participation

Degree of citizen power	*Citizen control* *Delegated power*	*Supporting independent community initiatives*	*Self-governing*	*Bargaining*	*Empower*
	Partnership	*Acting together*	*Facilitator*		*Partner*
Degree of tokenism	Placation	Deciding together	Stakeholder		Involve
	Consultation	Consultation		Consultative	Consult
	Informing	Information	Source of local knowledge		Inform
Non-participation	Therapy (education)				Influence
	Manipulation		Activist	Authoritative	
Key references	Arnstein (1969)	Wilcox (1994)	Vanderwal (1999)	Vanderwal (1999)	Robinson (2003)

community may be functionally illiterate. Social and institutional capital are some of the qualities that poor rural communities need significant support to develop. As a result their likely rate of 'demand-responsive development' will therefore (almost) always be slower than external agencies would like, particularly when they have budgetary and implementation targets to achieve. That means that there is often a tension between the quality of interventions and the scalability of approaches. More meaningful participatory processes – of the interactive type – are needed to increase the probability of sustainability of service delivery yet such processes take more time, money and require well-skilled staff to facilitate them. These conditions are often not available at scale.

Using these ladders of particicipation, we started the research by proposing three different approaches to community management which help enrich the general model. We categorise the sample into three broad typologies:

- Typology 1 – 'Direct Provision with Community Involvement': The community receives direct support on finance, materials and technical issues from an external agency and under that agency's control the community is partially involved in O&M. The external agency is usually the local government, a centralised public body or, on occasion, an NGO.
- Typology 2 – 'Community Management Plus': A community institution is responsible for O&M and service provision. This community institution remains voluntary and may not be legally recognised as the service provider but fulfils the role.
- Typology 3 – 'Professional Community-Based Management': The water system is operated by an authorised business-like organisation with a community institution either taking responsibility for service provision in a professional way or outsourcing this to other entities.

As illustrated in Figure 2.2, the intensity of community involvement is supposed to vary so that the degree of participation follows a normal distribution curve across the types. Based on that model, there is an implicit understanding that communities with higher average incomes are more likely to contribute through user charges with these funds enabling the professionalisation of the service with paid-for staff leading to the 'Professional Community-Based Management' consumer-orientated model. For example, the Government of Colombia Programa de Cultura Empresarial (business culture programme) supported programmes in which the government supports a professionalised community-based service provider that adopts good business practice, such as electronic book keeping, and hires paid-for staff to take on key aspects of O&M (Tamayo and García, 2006).

In communities with low average incomes and fragile livelihoods there is little additional capacity to contribute to the service, leading to a situation where 'Direct Provision with Community Involvement' emerges, with community members then ideally involved in key decisions over the service but not taking responsibility for the operations and maintenance of the service. Communities in the middle of these groups are more likely to participate in a conventional way through the

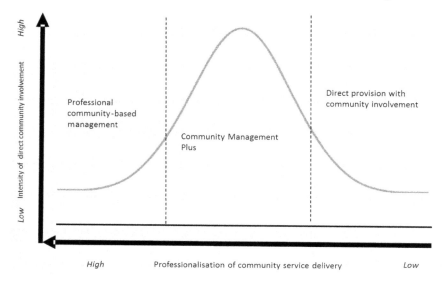

Figure 2.2 Level of community involvement across different typologies of community management

'Community Management Plus' model with members volunteering to take on key duties including operation, maintenance and administration as well as contributing modestly through user charges. Gram Vikas, the Indian NGO working largely in the state of Odisha, provides an example of community management plus. The NGO delivers an intensive preparatory period of capacity building and awareness-raising in villages before assisting with the construction of a new scheme. The community then takes responsibility for O&M with occasional call-down support provided by the NGO to the village water and sanitation committee (Thomas, 2013).

The above also means that community management must be seen in relation to other management models. Lockwood and Smits (2011) have produced a visual overview of different service delivery models, as presented in Figure 2.3. What this figure suggests is that there is a hierarchy in the professionalisation of service delivery models that range from self-supply – whereby households invest in and manage water services independently of any support (Sutton, 2008) – to what could be described as an urban-utility model – whereby a fully professionalised organisation manages services in a financially sustainable manner based largely on user fees. This study recognises this diversity but is focused exclusively on community management as it is the dominant and most common management approach in low and lower middle income countries for rural areas (Van den Broek and Brown, 2015). The research recognises that management models can overlap and that hybrid systems can develop. It is these conceptually blurred areas of overlap, particularly between direct local government providers and community management that this research will consider in more detail later in Part III of the book.

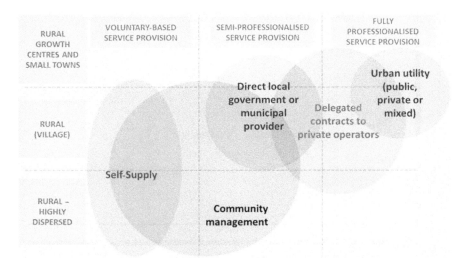

RURAL
GROWTH
CENTRES AND
SMALL TOWNS

VOLUNTARY-BASED
SERVICE PROVISION

SEMI-PROFESSIONALISED
SERVICE PROVISION

FULLY
PROFESSIONALISED
SERVICE PROVISION

**Urban utility
(public,
private or
mixed)**

**Direct local
government or
municipal
provider**

Delegated
contracts to
private operators

RURAL
(VILLAGE)

Self-Supply

RURAL –
HIGHLY
DISPERSED

**Community
management**

Figure 2.3 Service delivery models

The historical evolution of community management and support to it makes it important to differentiate between two layers of organisations that operate within the community management model. This includes the service provider level, which is the organisation(s) 'carrying out all the day-to-day tasks of operation, maintenance and administration of the water system', in the words of Hutchings et al. (2016, p. 27). They continue:

> Typically these tasks include: Operation: operating the engine of a pump, managing a treatment or disinfection facility; Maintenance: small preventive maintenance, like greasing of mechanical parts, cleaning of reservoirs, repairing leakages in the network and broken-down pumps and other corrective maintenance; and Administration: billing, tariff collection, book-keeping, reporting, governance.
>
> (Hutchings et al. 2016, p. 27)

Within an urban utility context the entity undertaking such tasks could be called an operator but as this research is focused on community management, the service provider will be referred to as the 'community service provider'. Under the Government of India (2013a) National Rural Drinking Water Programme (NRDWP) it is mandated that the community service provider should be a community committee know as a Village Water and Sanitation Committee, which will be discussed in more detail in the following chapter.

It is now widely argued that the community service provider should be recurrently supported and monitored by a range of other organisations (Bakalian and

Wakeman, 2009; Baumann, 2006; Kleemeier, 2000; Lockwood, 2002, 2004; Lockwood and Smits, 2011; Smits et al., 2013). These organisations will be known as the 'enabling support environment' that can provide support to and monitor the performance of the community service provider: 'The main objective of such support is to help communities in addressing issues they cannot reasonably solve on their own and gradually improve their performance in their service provider functions' (Smits et al., 2015, pp. 22–23). Although there is a strong consensus that this type of support is important to ensure sustainable and successful community management, there remains a lack of understanding about how to best structure and finance such support (McIntyre et al., 2014; Smits et al., 2011). This research is designed to fill such a knowledge gap which will be investigated further in the reminder of this literature review.

Providing effective and equitable services: household service levels

The discussion at the start of this chapter on the poor service delivery makes it necessary to further define what makes a water supply good or poor. That is an important conceptual discussion in order to assess whether community management leads to effective and equitable service delivery.

A water supply service is effective if water flows with certain agreed-upon characteristics. There is of course a difference in whether the water only trickles out or has a good pressure and flow, and whether the water meets quality standards or not. It is therefore important to define characteristics of the water service. The most commonly used ones are the quantity of water, its quality, the reliability and accessibility of supply. Taken together these make up a service level. For example, access to 50 litres per capita per day (lpcd) reflects a higher level of service than access to 25 lpcd. Some would argue that the costs or the affordability of the supply should be considered as part of the service level as well. While undoubtedly important, this is fundamentally different, as it is a reflection of the financial (or management) costs to get to a certain service level. It would often cost more to have access to 50 lpcd than to 25 lpcd. Costs are therefore not part of the service level definition itself, but reflect what is needed to reach a certain service level.

A differentiation needs to be made between:

1   normative service level (i.e. the level of service to which users are entitled, as per official norms or standards);
2   the design service level (referring to the specifications of the service level as per the design parameters of a particular system); and
3   actual service level, being the characteristics of the service that users actually receive.

The latter two may be above or below the norm. In the WASHCost project, the following ladder was developed for India (see Table 2.2), based on the NRDWP (National Rural Drinking Water Programme) guidelines. An acceptable service is

Table 2.2 Service ladder for India, as used in the WASHCost project

Service level	Quantity	Accessibility	Quality	Reliability/dependability
High	> 80 lpcd	0–10 minutes per day to collect water	Water quality has been tested independently using a water quality test kit.	As below, but a system for handling breakdowns exists and it functions well.
Improved	60–80 lpcd	10–20 minutes	Users are aware that rural water supply and sanitation officials have certified that there are no water quality problems.	A basic level and a system for handling breakdowns exists, but is not functional.
Basic	40–60 lpcd	20–30 minutes	No complaints from users.	Network supply according to an agreed schedule and duration. Handpumps are dependable but no system for handling breakdowns exists.
Limited	20–40 lpcd	30–60 minutes	Water is used for drinking but users complain of bad smell, taste, colour or appearance.	Network supply has scheduled times, duration and delivery but supply is haphazard. Handpumps are not dependable because recharge rates are low.
No service	< 20 lpcd	> 60 minutes	Water is unfit for drinking by humans or animals.	Network supply is haphazard. Handpumps are not dependable because groundwater is exhausted.

Source: Snehalatha et al. (2011).

thus one in which at least the normative service level is provided over time, and even possibly improved.

Sustainability in community management

Whereas the previous section discusses whether water services are effective at a certain point in time, it is also important to look at whether those services will be provided in the future. There are numerous interpretations of the concept of viability or 'sustainability' for rural water supply. Many organisations define sustainability as the maintenance of the perceived benefit of investment projects (including convenience, time savings, livelihoods or health improvements) after the end of the active period of implementation. Building on the definition by Abrams (1998), who describe sustainability as 'whether or not something continues to work over time', it could be defined as whether water continues to flow over time.

Whereas that is a simple definition, it is one that can only be assessed with the benefit of hindsight. Different frameworks have emerged to both assess whether water indeed has kept on flowing and to predict whether water is likely to keep on flowing in the future as well. In their review of five recently-developed sustainability frameworks, Boulenouar et al. (2013) found that many of the recent frameworks looked at parameters at community level, but also – to a lesser extent – into parameters at higher institutional levels, including districts and even national level. The mentioned groups of indicators are scored at these institutional levels, so as to come to an overall assessment of the likelihood of good rural water services. In a way, these tools take a snapshot of the degree to which the factors are in place and, based on that, identify a likelihood of service sustainability.

This book makes a particularly strong connection between finance and sustainability as a key factor that needs to be addressed within the community management model. The first step to understanding financial viability is to determine the costs. For this study, we use the cost categories that follow the pattern of service development and delivery, as defined in Table 2.3. A variation on the approach developed by the WASHCost project was adopted, using the components defined in the table.

Next to the costs, we need to understand how these are paid for. The UN-Water Global Analysis and Assessment of Sanitation and Drinking Water report makes clear that there are three sustainable sources of finance for water services. These are known as the 3Ts (tariffs, taxes and transfers): '(a) the monies paid by the users of the services ("tariffs"), (b) the monies provided by domestic taxpayers through governments ("taxes"), and (c) the monies provided by foreign countries ("transfers")' (UN Water and WHO, 2014). Getting the right mix of these 3Ts is of strategic importance in terms of delivering sustainable services (OECD, 2009), yet it remains one of the biggest challenges (Burr and Fonseca, 2013; Hutton et al., 2007; Hutton and Varughese, 2016). At the global level it is estimated that capital financing of $114 billion per annum would be required to achieve universal access to water supply, which is three times the current investment levels (Hutton and Varughese, 2016).[2] However, the greater challenge relates to the growing operation and maintenance costs (Baumann, 2006; Briscoe and Malik, 2005), with these estimated to become a larger annual cost than capital investment during the 15-year period between 2015 and 2030 (Hutton and Varughese, 2016). They are set to rise from around $18 billion to $128.8 billion over that period (ibid.). In this context, a basic tenet of the community management model has conventionally been that user tariffs should cover the operation and maintenance costs of rural water supplies and users should contribute 10 per cent of the capital costs (Joshi, 2003). Taxes or transfers should then cover the remaining 90 per cent of the capital costs

Table 2.3 Definitions of cost categories

Cost category	Description
CAPITAL OR 'ONE-OFF' INVESTMENT COSTS	
Capital expenditures – hardware and software (CapEx)	The capital invested in constructing fixed assets such as concrete structures, pumps and pipes. Investments in fixed assets are occasional and 'lumpy', and include the costs of initial construction and system extension, enhancement and augmentation (also called CapEx on hardware), as well as one-off work with stakeholders prior to construction or implementation, extension, enhancement and augmentation, such as costs of one-off capacity building (called CapEx on software).
RECURRENT OR ANNUAL COSTS	
Operating and minor maintenance expenditures (OpEx)	Expenditure on labour, fuel, chemicals, materials, regular purchases of any bulk water. Minor maintenance is routine maintenance needed to keep systems running at peak performance, but does not include major repairs.
Operation expenditure enabling support (OpEx Enabling Support)★	This includes expenditure on support activities direct to local-level stakeholders, users or user groups, such as support to service providers and ensuring that local government staff have the capacities and resources to help communities when systems break down or to monitor performance. It could also include elements of macro-level support, planning and policy-making that contributes to the service environment, but is not particular to any programme or project. Indirect support costs include government macro-level planning and policy-making, developing and maintaining frameworks and institutional arrangements, and capacity building for professionals and technicians. However, these are usually hard to define.
Capital maintenance expenditure (CapManEx)	Expenditure on asset renewal, replacement and rehabilitation costs, based upon serviceability and risk criteria. CapManEx covers the work that goes beyond routine maintenance to repair and replace equipment in order to keep systems running. Accounting rules may guide or govern what is included under capital maintenance, and the extent to which broad equivalence is achieved between charges for depreciation and expenditure on capital maintenance.

Note: ★This category has been adapted from the Expenditure on direct support (OpexpDS) and Expenditure on indirect support (OpexpIDS) in the life-cycle costing approach. This is because during the WASHCost research there was extremely limited data available on OpexpIDS and therefore it becomes useful to consolidate it together into an overall OpEx Enabling Support category.

Source: adapted from Fonseca et al. (2011).

with the belief that as countries get wealthier (like India has been getting in recent decades) a greater proportion of capital costs will be covered by domestic tax (Franceys and Cavill, 2011). Yet research has shown that user tariffs rarely cover even basic operational costs and even more rarely cover the full recurrent costs including capital maintenance (Burr and Fonseca, 2013; Moriarty et al., 2013; Ratna Reddy et al., 2010).

The described situation points to a fundamental flaw in the logic of the demand-responsive approach to community management – that rural people in low and lower middle income contexts will consistently have the 'willingness-to-pay' for water services at a sustainable level. The rationality of focusing on willingness-to-pay – as a proxy for the concept of 'demand' – can be summarised as follows:

> If people are willing to pay for the full costs of a particular service, then it is clear indication that the service is valued (and therefore most likely to be used and maintained) and that it will be possible to sustain and even replicate the project.
>
> (Whittington et al., 1990, p. 294)

However, as stated above, willingness-to-pay is variable between and within communities and is often not at the required level (Marks and Davis, 2012). There is evidence that there is a threshold effect in terms of effective demand with service levels having to be sufficiently high for users to pay (Fonseca, 2014; Koehler et al., 2015). Moreover, there is recent evidence that communities have a lower willingness to pay for community-managed services than other service delivery option (Hope, 2015). Even if there is some evidence that rural water service users are willing to pay a tariff at some level, the most fundamental problem is that it rarely covers the actual operation and maintenance costs of supply leading to sustainability problems in the service delivery cycle (Franceys et al., 2016; Franceys and Cavill, 2011; Marks and Davis, 2012).

Replicability

As well as effective and sustainable services, a key focus of this book is about community management programmes that are replicable across scale. This refers to the possibility to apply a certain approach to community management at a large number of localities. In the past, highly participatory and empowering approaches have been developed to rural water supply, but those often required high levels of external, social development or NGO-type input, for example, highly skilled facilitators who would spend lots of time working with communities. Such approaches were often not scalable, because the human and financial resources were not available. That is not only an issue of efficiency, but of sheer availability of the resources. Moreover, such approaches were often dependent on a relatively high degree of capacity within the recipient community (e.g. in the form of a few exceptional leaders or otherwise extraordinary circumstances). Or to turn it around, an approach is replicable if it is reasonably efficient (i.e. does not require large amounts of resources to achieve a certain level of services) and if it works well in average conditions, with averagely skilled support entities and averagely performing recipient communities. For this study, we focus mainly on those community-management programmes that have already achieved a certain level of scale, to avoid selecting only those with exceptional skills or leadership.

Chapter summary

This introductory chapter has set this book in context. It has provided an overview of the history of community management, making links with the broader movement of participatory development. In doing so, it has presented some of the critical challenges facing the community management model and developed a series of typologies that explain the different forms of the model. It then ended by clarifying what the researchers understand to be a good rural water service – one that responses to the challenges of effectiveness, sustainability and replicability.

Notes

1   An improved drinking-water source is used as the basic measure of access in the monitoring programmes for the MDGs. It is 'defined as one that, by nature of its construction or through active intervention, is protected from outside contamination, in particular from contamination with faecal matter' (WHO and UNICEF, 2013).
2   All dollar values in this book refer to US dollars.

3 A systematic review of success factors in the community management of rural water supplies over the past 30 years

Introduction

Having seen the background to community management and some of the challenges with it, this chapter seeks to systematically assess the characteristics of success in community management over the past 30 years around the world so as to further develop our understanding of what successful community management looks like. It focuses specifically on experiences outside India, given that the bulk of the remainder of this book is focused on India only. Through this process it will also assess a more fundamental hypothesis: that success in the community management of rural water supply is directly related to broader socio-economic trends in a country. Building this argument, we point to the historical evidence from high-income countries which suggests a correlation between economic growth and the expansion of water services (Gerlach and Franceys, 2009). There are a number of reasons why this trend may also be apparent in terms of the community management of rural water supply. Withstanding high rates of economic inequality, growth in gross domestic product (GDP) per capita is likely to lead to higher levels of domestic financial resources that users and government are able to contribute to the water supply services. Specifically, the chapter builds on that hypothesis and the broader idea of community management plus, as outlined in the previous chapter, to address the following two questions: what 'plus' factors are associated with successful community-managed rural water supplies? And is the socio-economic setting indicative of the likely success of a community-managed rural water supply? Through this investigation the chapter intends to provide a strong evidence base for policy-makers and researchers on the type and nature of support that is required if community management plus is to be a success as well as indicative findings on the relationship between broader socio-economic conditions and the success of community management.

Method and approach

Recognising the breadth of studies on community management in rural water over the past 30 years, especially in the grey literature, a systematic review with meta-analysis was selected as an appropriate approach for comprehensively synthesising the available evidence (Petticrew and Roberts, 2006).[1] Initially a total of 2,544 potential case studies were found in the different sources reviewed. To determine whether the cases were to be selected for further analysis they had to meet four basic criteria:

1   located in developing countries;
2   located in rural areas;
3   systems managed partially or entirely by the community; and
4   systems functioning and delivering water to the community (at the time of the case study).

If these were not met, or if insufficient information was provided on these points, they were excluded from the analysis. We considered both academic publications and grey literature sources. As shown in Table 3.1 below, through this initial selection process 174 cases were selected as the primary sample from which we were to conduct further categorisation and analysis.[2]

The case studies identified as successful in the first phase were then analysed using the 'Success Framework', a qualitative tool created for this study which included information about up to 41 different aspects of the case study, including scale, scope, technology, service levels, population characteristics, and so on. Using the collected information, and noting that all case studies had already been selected on the basis that they were at least a marginal success, a score between 0 and 5 was allocated, with 0 representing a marginal success (e.g. a water supply that delivers water to the community, but is not well managed nor provides good service levels) and 5 representing a full success (e.g. continuous delivery of water and the characteristics of a well-managed system).

Using the information from the Success Framework database, the scoring of the case studies was undertaken using a 'Scoring Sheet' based on six different aspects of water service delivery. These were based on 'EEVERT' assessment principles (Franceys, 2001), which are defined as follows:

- *Effectiveness*: is it working/delivering water?
- *Equitability*: can all benefit, particularly the poorest?
- *Viability*: will it continue to deliver? In financial, technical and environmental terms?

Table 3.1 Number of case studies by region

Region	Number of cases
Latin America and the Caribbean	18
Middle East, North Africa, Afghanistan and Pakistan	21
Sub-Saharan Africa	79
Developing Asia	56
Total	174

- *Efficiency*: is it being achieved with optimum use of resources?
- *Replicability*: can it be repeated by others, can it be 'scaled up'?
- *Transparency*: is it apparent/understandable to all how it works? Is there communication between the service provider and the community?

After scoring each aspect individually, an average of all was taken as the final score for the case study. The scores were peer-reviewed. The scores were then plotted against the GDP data and analysed visually and using basic correlation analysis. For this purpose, the case studies were grouped into either regions or, when data allowed, specific nations. In this chapter that analysis is shown for the Latin America and Caribbean and Sub-Saharan African regions only.

Description of the sample

This section provides a descriptive overview of the sample. It helps clarify the characteristics of successful community management schemes and identifies key themes that emerge from reading the case studies. Across the sample, 50 per cent of schemes provided water via some form of public stand-post while 28 per cent supported handpumps and 22 per cent provided household piped water supply. We also identified the number of cases of the different types of community management, as presented in Table 3.2.

In order to focus down on the 'plus' factors that have worked in the most successful cases, we then restricted the analysis to cases which scored above 3 in the success ranking. This process left 72 cases in total. This new sample of 72 cases were then analysed in more detail for evidence of the *plus* factors that had contributed to their high levels of success. This analysis indicated that it was necessary to consider plus factors in terms of internal and external factors.

Identifying internal plus factors

Focusing on the 72 high-performing case studies, an analysis was conducted to identify the most common internal plus factors found in these successful cases. This led to the identification of three broad themes that were classified as 'collective initiative', 'strong leadership' and 'institutional transparency', which were deemed to be influential success factors across the case studies. Collective initiative was evidenced through a variety of factors including a communal ethos of self-help and

Table 3.2 Number of case studies by community management type

Region	Number of cases
Typology 1 'Direct Provision with Community Involvement'	28
Typology 2 'Community Management Plus'	122
Typology 3 'Professional Community-Based Management'	39

responsibility, equitable participation in decision-making from across the community including women and disadvantaged groups, and a notion of shared ownership of the scheme. Strong leadership included cases when exceptional individuals or groups of individuals from the community have been able to provide supervision, monitoring and evaluation of systems and workers, as well as take the role of everyday and strategic decision makers. Institutional transparency relates to cases where accountability mechanisms are built into community institutions responsible for water supply, including democratic procedures and the disclosure of financial and other performance data. These traits were neither mutually exclusive nor necessarily apparent in all cases of success but rather were associated with certain types of schemes and different stages of the service delivery cycle.

Following a categorisation of case studies, these three factors were then evaluated against the length of functioning of case studies. An emphasis was placed on identifying the single most influential internal plus factor for each case study then assessing this distribution against the length of project. It was found that collective initiative was the key internal plus factor for schemes that have lasted from zero to five years, but that in schemes that had lasted more than five years the distribution shifted to become more balanced across the categories. This indicates that while high initiative of the community is vital for the start of the project, in order to sustain it, a more balanced approach with strong leadership and clear transparency is likely to be needed. Wider evidence shows that the bottleneck for failure in community management is likely to come some years after a project ends when operation and maintenance fails or capital maintenance is not fulfilled (Baumann, 2006; Harvey and Reed, 2006; Lockwood and Smits, 2011). To avoid this, community organisations need to demonstrate to the community the basis behind setting appropriate user charges or arrangements for adequate funding while showing clearly how this relatively large amount of money will be spent. Transparency is therefore so important for longevity because if users have doubts about the system, user charge collection systems are likely to fail. As one may expect, the review also suggests the presence of strong leadership with the right skills to manage the overall operation including human resources, management and finance, will also help drive the community in delivering sustainable services.

The three internal plus themes were also evaluated against management typology. Community initiative formed the building block behind the community management plus model, yet for this model to evolve to the next stage as professional community-based management, transparency and leadership are needed as well. Perhaps surprisingly, a key factor for the direct provision model is institutional transparency. We found that in many cases, under this typology, community-based organisations were responsible for O&M using funds from user charges collection. For this purpose the community-based organisations were required to report the progress of the project, especially the disclosure of the financial status to the project donors, who were subsidising the operation. This may explain why transparency was so important with the direct provision with community involvement model.

Identifying external plus factors

Each one of the 72 most successful case studies were also analysed for external plus factors, such as the services provided by external agencies in support of community management. Unsurprisingly, finance played a critical role. As shown in Figure 3.1, it was found that over 90 per cent of these high-performing cases received external financial support for different expenditures (capital expenditure, operating expenditure and capital maintenance expenditure) and provision of materials from external organisations. Eight per cent of cases had access to loans and microfinance, and these cases were found particularly in Asian countries like Pakistan (Asian Development Bank, 2008; Padawangi, 2010) and China (World Bank, 2002). Capacity building for management was provided to more than half of cases and around one-third could seek advice on technical issues related to O&M from the external organisation. In total eight main forms of external support were identified (see Figure 3.1).

In order to determine the importance of these external plus factors on the service sustainability, the eight forms were evaluated based on the reported length of operations. Based on the results, external factors such as 'financial support and provision of materials' and 'capacity building for management' were found equally important in both younger and older systems. However, cases which were less than five years from initiation had more access to 'advice on management and finance', 'loan and microfinance', 'supply chain of spare parts and services' and 'capacity building on technical skills'. The level of external support to younger schemes may

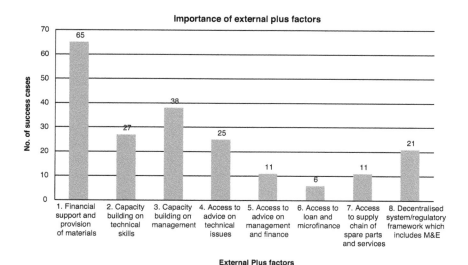

Figure 3.1 Importance of external plus factors in most successful cases

be reflective of the project cycles of many programmes, when funding and support mechanisms from implementing agencies are still available. Lastly, cases more than five years in length were more likely to report a 'decentralised system/regulatory framework'. This result indicates that the presence of governmental support through a decentralised system and reformed policies are instrumental in sustaining schemes over the longer term.

The eight characteristics were also evaluated based on management typology. No advice on management, access to microfinance and loan or to supply chain were observed in the direct provision with community involvement model, but were observed in the other two community management types. This is likely due to the fact that external organisations tend to be in charge of most parts of the system, therefore the need to provide support to the communities is seen as less important. Community management plus appears to require various forms of external support, from financial support to advice on technical/management issues, while professional community-based management required a broader type of enabling support (e.g. access to loans and regulatory framework). The importance of this type of broader enabling environment clearly increases when the community starts managing the water system in a more professionalised and legalised way.

The socio-economic setting of success

In testing whether the socio-economic setting is indicative of success in community-managed rural water supply, we plotted the success scores from all the 174 case studies against GDP (PPP) per capita for respective regions and countries. The reported cases of success were more common after 1995 so the graphs emphasise this time period. It is unclear whether this reflects a growth in success since 1995 or the availability of published case studies from this time period. In Latin America and the Caribbean, a region with a broadly middle-income population, the data shows that the level of performance in community management has improved alongside economic growth (see Figure 3.2). Based on the working hypothesis, we suggest this is because internal financial resources have grown within communities, improving cost recovery. This is supported by reading the case studies, with all Latin American cases lasting five or more years, showing evidence of consistent financial contributions from users. Similarly, among most successful cases, with scores above 3 (Davis et al., 2009; Madrigal-Ballestero et al., 2013; Prokopy, 2009; Whittington et al., 2009; World Bank, 2001a), four were professional community-based management with trained staff dedicated to the operation and maintenance of the systems and/or efficient and transparent administration. We believe such models become more common and effective as broader trends of socio-economic development occur, particular in terms of the (equitable) education of the population.

While an indicative relationship between the socio-economic indicator and the success of case studies was visible in other regions, for Sub-Saharan Africa there appeared to be no discernible relationship (see Figure 3.3). Even when accounting for programmes evaluated more than five years after initiation, the scores were

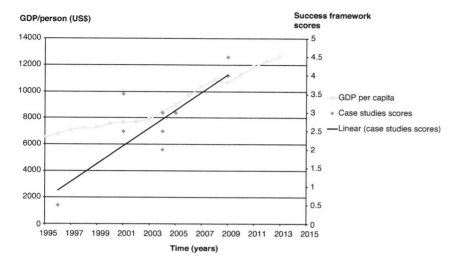

Figure 3.2 Success scores versus GDP per person in Latin America and the Caribbean

equally distributed around the mean score of 2.5 with a range of scores observed for every year, including the most recent. In part this may be explained by the high heterogeneity of the region where countries have significant differences across economies (from farming to oil production), development stages and political

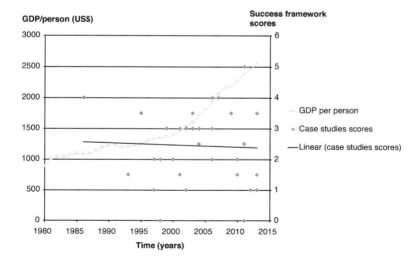

Figure 3.3 Success versus GDP per person in Sub-Saharan Africa

stability. However, the high variability of the scores over time could also be explained by the dependence on international aid observed in the case studies. In Sub-Saharan Africa, case studies were more likely to report an absence of internal financial resources or problems with tariff-setting and bill collection, while external support from government sources was not as common. Instead, external aid was more common and so this dependence on external support could explain why Sub-Saharan Africa is not following the economic trend.

The fuzziness of the data presented a challenge in terms of statistical testing of the hypothesis, however the basic correlation between GDP per capita and the success scores was examined using a simple linear regression for each of the major regions presented in this section. The results presented in Table 3.3 show that in Latin America there is a strong level of correlation between higher levels of GDP per capita and higher success scores. However, in Sub-Saharan Africa, there was no correlation between the variables. At higher levels of GDP, the evidence supports the hypothesis that on a regional or national scale the underlying level of wealth of a society is a predictor of success in community-managed rural water supply schemes. However, at lower levels of GDP, as found in Sub-Saharan Africa, this relationship does not hold. This is likely to be explained by the high levels of international aid found in the water sectors of low-income countries, which is distorting the relationship between success and wealth.

Discussion and implications

Research had previously emphasised the importance of internal community characteristics, such as social cohesion, as determining factors for success yet the community management plus literature has predominantly been focused on the institutional mechanisms for providing support down to communities (Schouten and Moriarty, 2003; Baumann, 2006; Lockwood and Smits, 2011; Moriarty et al., 2013). While the analysis presented in this chapter affirms the critical role of such external support, it re-emphasises the importance of local context. This is illustrated by locating the relative importance of internal plus factors against the length of operation of systems. Collective initiative is particular important at the start of community management yet institutional transparency and leadership are the key internal characteristics for longer programmes. This has implications for authorities

Table 3.3 Correlation between GDP per capita and level of success in community-managed rural water supply

	Latin America and the Caribbean	*Sub-Saharan Africa*
Correlation coefficient	0.800	0.063
R-squared	0.640	0.004
p-value	0.005	0.742

designing community management training programmes, with social mobilisation only a starting point that should be accompanied by periodic support that focuses on leadership development and administrative processes within community water institutions.

The review also continues to support the requirement for external support provided down to the community from a number of entities that operate in a broad 'enabling support environment'. This sphere of support should be directly built from the higher echelons of state governments and international institutions down to local governments and other local entities. In terms of what type of support is required at the community level, a number of basic components were visible in successful cases. Over 90 per cent of high-performing cases involve financial support to the community, while technical advice and managerial advice are also important. Over the longer term more complex forms of support are required, which can be broadly defined as a regulatory framework. This framework of government policies and standards is particularly critical as the wealth of populations increase and communities move along the ladder from simplistic management models to the professional management approaches found in places such as Latin America.

This chapter implies that success in community management depends on the coming together of three interrelated components: internal plus, external plus and underlying socio-economic wealth. Depending on a range of factors, there may be sufficient community capacity to support certain elements of this input for some time, yet it is highly unlikely to be adequate to fulfil all these types of inputs, especially over the long term. Crudely speaking, low levels of 'internal plus' mean that a high degree of 'external plus' is required while high levels of 'internal plus' mean a community institution will be less dependent on 'external plus', although still in need of support at times. However, the capacity to provide either internal or external plus is directly related to the prevailing socio-economic wealth in a society. Increased national wealth leads to expansions in public spending on water supply, while household wealth results in improved payments of user charges, and overall wealth is both resultant from and a driver of institutional and economic capital throughout the population. When these conditions exist, as in some parts of Latin America – and as will be elaborated in the remainder of the book, also in India – community management becomes more professionalised and can deliver good and lasting services. Yet, in its absence, such as is the case in large parts of Sub-Saharan Africa, community management struggles to provide long-lasting services.

Chapter conclusions

Unacceptable failures in rural water service delivery have called into question the prominence of community management as the dominant service delivery model in the sector and yet, at the same time, community management has played a significant role in the expansion of water services to rural populations around the world in recent decades. Bringing together what initially appear contradictory statements, this chapter identifies what has worked in terms of community management over the past 30 years. It shows that when community management is successful a number of internal and external elements come together to create what has been classified as a 'plus' to the standard community management model. The research links the presence of these internal and external success factors with the broad levels of socio-economic development found in populations. In Latin American and the Caribbean, the growth in resources available at either the household or government level appears to be leading to an improvement in the overall levels of success in the community management model. However, in the lower-income region of Sub-Saharan Africa, no relationship was observable between the success of community management and the level of GDP per capita. This was thought to be explained by the high levels of international subsidy channelled into projects across Sub-Saharan Africa that distorts the relationship between success in rural water supply and national wealth.

Acknowledgements

This global overview chapter is based on the author's research paper: P. Hutchings, M. Y. Chan, L. Cuadrado, F. Ezbakhe, B. Mesa, C. Tamekawa and R. Franceys (2015) 'A systematic review of success factors in the community management of rural water supplies over the past 30 years', *Water Policy*, 17(5), 963 (IWA Publishing).

Notes

1   This section provides an abridged version of the methodology. For the full systematic review methodology please see Hutchings et al. (2015).
2   All these documents have been uploaded to the virtual reference database Mendeley. We encourage interested parties to make use of the database. It is available at www.mendeley.com/groups/4499771/rural-water-community-management-plus.

4 Revisiting the history, concepts and typologies of community management for rural drinking water supply in India

This chapter investigates the research context of community management in India. The first half of the chapter focuses on the story of community management within India before the chapter defocuses to consider the contemporary operating context for rural water services. This includes an overview of the governance arrangements found in India and an analysis and discussion about the different levels of development found across the country.

The history of community management in India

India has a long history of community management. Early experiments were tried as far back as 1964, with the World Health Organization and UNICEF Banki and Mohkampur projects in Uttar Pradesh running with some limited success until 1994 and 1976 respectively, and the sister Pharenda project reported as still on-going at the time of the last citation (WSP, 2002). Tracing the genealogy of community management from these early initiatives to the present day, this section begins by drawing on James's (2004, 2011) synthesis reports to identify four broad categories of community management initiatives. These include independent cases where communities have simply taken complete charge of water supply when government services have failed, small-scale NGO initiatives, larger scale donor–NGO schemes support by bilateral and multilateral agencies, and then the post-sector reform government-supported programmes that emerged from 1999 onwards. The learning from each category will be briefly discussed in order to demonstrate how community management has changed throughout this period.

With the provision of safe drinking water constitutionally mandated as the government's responsibility it is rare for communities to be completely autonomous in the management of drinking water. However, there are limited cases of reportedly independent piped schemes such as the case in Kolhapur (Maharashtra) that ran from 1979 until the 1990s (James, 2004). This occurred after the District Administration refused to take on a government-constructed piped water network, so taking their own initiative community members from four villages came together to form an unofficial committee that took responsibility for the piped network. Without any further support they managed the scheme for 20 years, even creating a big surplus in the committee's accounts (ibid.). While this case of unsupported community-managed piped water supply is relatively rare, it does demonstrate that it is possible. Yet its eventual failure also highlights how 'even a successful community management initiative requires a support structure to cope with external shocks and stresses' (ibid., p. 39).

As opposed to the paucity of completely independent cases, there have been many *small-scale NGO projects* that have been significant in developing the contemporary practices of community management in India, as James (2004) illustrates with a number of examples from Gujarat, Maharahstra, Karnataka and other states. These examples merely touch the surface of the numerous NGO programmes operating over the past decades yet they highlight the importance of NGOs in the establishment of community management. Yet, although there were examples of success, arriving at such outcomes often involved time-consuming 'trial and error-based experimental' approaches that were often costly in terms of resources and which require specialist skill sets (ibid., p. 49). This makes this kind of approach only limitedly scalable – as such resources and skill sets are often not available, and cannot easily be employed in programmes that cover many villages.

Throughout the 1990s and the early 2000s, the demand responsive approach (DRA) to community management was introduced to the country through a number of bilateral and multilateral donor–NGO programmes (Black and Talbot, 2004). In India as elsewhere, an actual DRA has been implemented only to a limited extent. The 10 per cent upfront contribution – through which users are supposed to express their demand – is waived in many cases or – as much anecdotal evidence suggests – paid for by contractors, so they can get on with the work. The 100 per cent recurrent cost recovery principle has also not been applied systematically. The myriad of direct payment of water-related costs by public bodies, such as the energy costs for pumping, the costs of water quality testing and major repairs, means that actually only a small part of the costs have to be paid by users: minor repairs and maintenance and some of the operating costs, like salaries of pump operators.

Notwithstanding these limitations, the concept of the DRA has been extremely widespread in internationally supported programmes with examples including the Kreditanstalt Für Wiederaufbau (German Development Bank) funded Aapni Yojna Project in Rajasthan (1994–2004), World Bank programmes in Maharashtra and Karnataka (1991–2000), and the World Bank Swajal Project in Uttar Pradesh (1990s). As opposed to the smaller-scale NGO approaches, these initiatives had budgets of $60–100 million and sought to serve a larger number of villages (500–1,000), often making use of smaller NGOs as partners (James, 2004). Professional approaches to community management were developed in this period, including building participatory methods into the design stage of programmes, scheduled training schemes with community members to build capacity, and tripartite agreements between village water and sanitation committees (VWSCs), support organisations and overall programme managers. However, despite professional practice, government requirements in areas such as procurement prevented community management flourishing beyond these programmes as community entities were unable to make use of allocated government funds or access government procurement processes (James, 2004, 2011).

A new form of government-supported community management emerged from 1999 onwards. In that year, the Government of India implemented sector reform pilot projects (SRPPs) in 67 districts across 26 states and so began the process of integrating community management into its national policy. In many states, new institutions were formed, including district-level water and sanitation committees, which received funds directly from the union government, by passing state level agencies. While there was some success in the pilot programmes, there was also resistance to change from officials who were used to a supply-driven model along with inadequate support at state and district level 'to provide backstopping and trouble-shooting' when initiatives failed (James, 2004, p. 39).

Despite these flaws, in 2002, the Government of India launched the Swajaldhara programme. The Swajaldhara programme advocates community management along the following principles: a demand-driven approach; village-level capacity building for community management through VWSCs; an integrated service delivery mechanism that streamlined the functioning of the government agencies involved; demand-responsive approach based on cost-sharing by users (100 per cent of operation and maintenance costs; 10 per cent of capital costs); and, water conservation measures through rainwater harvesting and groundwater recharge measures (Government of India, 2003). In practice it is questionable whether these claims were met even in successful schemes, as the Swajaldhara claim of 100 per cent operation and maintenance covered by the community does not reflect the many indirect (and hidden) subsidies in India that support rural drinking water supply, including electricity subsidy and administrative support to VWSC through the local government system. There are also cases when the programme has been poorly implemented, such as the one highlighted by Srivastava (2012), where the Swajaldhara programme in one area merely became a sham with no community management but a water supply run by local elites for their own benefit.

The sheer scale of the reform meant institutions at many levels did not have the capacity to implement the aspirational objectives and the Swajaldhara programme 'had roughly the same impact on sustainability as the regular … supply-driven model followed in the country since 1972–1973 … largely because of the inadequate preparation and capacity building – especially among the engineers as well as the community and NGOs' (James, 2011, p. 54). Notwithstanding the criticism, Swajaldhara was still significant as it legalised community management within the prevailing governance model, providing a formally recognised legal basis for communities to become service providers and thus removed barriers regarding their access to government funds and procurement procedures. Perhaps most significantly the Swajaldhara provided an impetus for a number of highly successful state-based programmes to flourish in the last decade including the Water and Sanitation Management Organisation (WASMO) in Gujarat, Jal Nirmal in Karnataka, Jalanidhi in Kerala and Jalswarajya in Maharashtra (James, 2011; Lockwood and Smits, 2011).

Following Swajaldhara, India is now home to a rich diversity of community management experiences. However the latest policy programme from the Government of India has sought to further formalise the model within the broader system of local self-government. Launched in 2009, the National Rural Drinking Water Programme (NRDWP), the successor to Swajaldhara, has consolidated the importance of the local government Gram Panchayat institution in rural water supply with greater responsibility and funds devolved to this level. The Gram Panchayat is the lowest level of government in the Panchayat Raj system of government that operates in rural India. These bodies have an elected president covering a 'village' (though typically each village covers a number of habitations) and are responsible for many public services, including domestic water supply. However, under these guidelines, a VWSC is formed and operates as a sub-committee of the local self-government (Government of India, 2012a). These nominated committees have between 6 and 12 members including the President of the Gram Panchayat and with a quota of at least 50 per cent representation of women, and are charged with the administration, operation and minor maintenance of rural water supply. However, the close institutional relationship with the Gram Panchayat means the VWSC is far from autonomous.

Evidence from another study (Rout, 2014) and as confirmed in the case study chapters later in this book indicates that this has often lead to dual systems developing whereby in certain villages the Gram Panchayat simply becomes the direct service provider while in other villages the VWSC are formed to enact community management with support from the Panchayat institutions (Rout, 2014). In many ways, the NRDWP promotes an institutional structure that is both robust and admirably malleable in that various institutional variations can emerge, even within the same programme. Yet this can also mean a lack of clarity over the exact nature of institutional arrangements, leading to questions over who takes key roles such as service provision or service monitoring. Furthermore, the diversity of approaches to rural drinking water supply is likely to grow further as the NRDWP comes to the end of its five-year cycle while the Government of India has abandoned its Planning Commission, replacing it with the NITI Aayog (National Institutions for Transforming India) which has less direct power over policy. Hence, in the long term, this move is likely to mean more freedom for states' governments to promote different models for rural drinking water supply (Government of India, 2015a). Together, this historic review shows that there are many different types of community management, and current policy trends are likely to lead to even greater diversity in practice. Yet the range of models and changes in the policy landscape now mean there are tensions – or at least conceptual uncertainties – with regards to the role of communities vis-à-vis the state.

Different typologies of community management across India

Based on an extensive review of grey and academic literature, as well as two expert stakeholder consultation meetings held in Hyderabad and Delhi during August and September 2013 respectively, we identified a sample of over 90 programmes that followed a community management approach and that were reported as successful, in the sense that the water supplies managed by these communities were generally providing adequate services with a meaningful role for communities. Based on the scale of operations described in the respective reports, we can estimate that these programmes cover 31,693 villages out of an all India total of 597,483 villages (Census of India, 2011a). Taking the average population size per village of 1,395 people, this suggests that at least, approximately, 45 million out of the 830 million rural population, or over 5 per cent, are receiving reportedly successful community-managed rural water services. It is expected that there are many additional cases that are not reported in the literature and therefore in that sample. However, this initial analysis still indicates that community management represents a viable model for a significant minority of people but that the majority of villages in India are not following a successful community management model, because either community management is not successful or more traditional top-down models are followed.

Making an assessment on the management model based on the typologies discussed in Chapter 2, 68 out of 92 cases were found to contain enough information to classify the case studies into the typologies. Four of the larger scale programmes were classified in each typology as the exact institutional arrangements appeared to vary across the programmes. This follows the pattern reported by Rout (2014) in her analysis in Odisha, which found both a form of direct provision and community management in the same programme. Beyond these multi-classified programmes, around one-quarter of the cases were characterised by direct provision with community involvement. Direct provision with community involvement by the GP is extremely common in India and is expected to account for many more programmes across the country than reported here. The ones included here are those which have a dedicated community engagement initiative alongside the direct provision and which were identified in our initial review. The distinction between the decentralised direct provision by the GP, that in its pure form is not a form of community management, and a model where the GP remains the service provider but is actively supported by a community body, such as a water committee, is where we consider the line to be between community management and government provision. Beyond direct provision, over 60 per cent of the programmes were classified as community management with support. This form of community management reflects the traits most commonly articulated to describe the model, with a community entity taking the role of service provider with support from other entities. Finally, the most advanced form, professionalised community-based management, was common in around 14 per cent of the cases. The development of these mini-utility type operations governed by community institutions is expected to grow when (and if) the proceeds from India's economic growth spreads into the rural areas, especially in areas close to urban centres. The snapshot provided in Figure 4.1 gives our best estimate at the current state of play in India but we expect this is likely to change. In fact, we contend that it must change as India develops if community management is to remain an integral part of its water policy.

This initial analysis presented in this chapter has shown that community management is a long-established phenomenon in the rural drinking water supply sector that has been the subject of critical discussion for a number of years. Yet

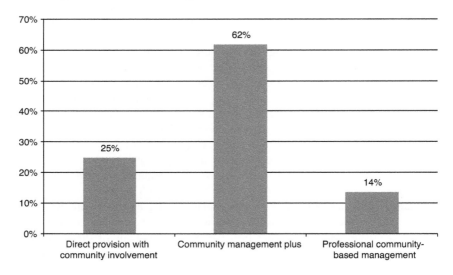

Figure 4.1 Community management in India by typology

Note: Calculated from Community Water Plus sampling frame.

there is no single Indian community management model but rather a loose and overlapping collection of models with varying degrees of community involvement and external support. As one can expect, there is no 'one size fits all' approach and so understanding that communities will need different types of support, which will likely be relational to their own internal carrying capacity, governments are in a better position to successfully take up the role of a successful support entity if they can better differentiate the types of support needed in different situations.

Setting community management within the broader context of rural water supply in India

This chapter now defocuses from the specific challenge of community management to place the previous discussion within the broader context of the Indian operating context. India has an improved water source coverage rate of 96 per cent in rural areas showing that it is moving towards universal access (WHO and UNICEF, 2013). Analysis shows that this is above the international trend line for countries with a similar GDP per capita level, as discussed later in this book. However, the Census of India (2011b) also provides data by the population living with 'piped-on-premises', also known as household or yard connections. This figure is at the much lower level of 31 per cent (ibid.), which is below the international trend line when compared with countries with similar wealth.

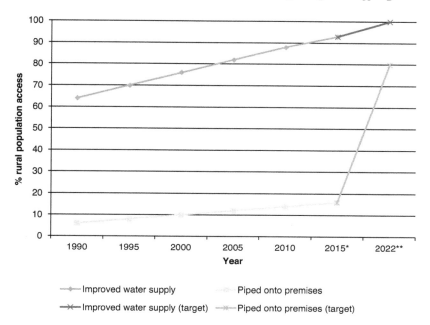

Figure 4.2 Rural water supply coverage rates and targets from Government of India

Sources: data from Census of India (2011b) and WHO and UNICEF (2015); targets from Government of India (2013a).

This is significant as the Government of India has recently set ambitious new policy targets to 'ensure at least 80 per cent of rural households have piped water supply with a household tap connection' by 2022 (Government of India, 2013a, p. 2). This means that it plans to expand access to such facilities to around 400 million people in less than a decade. As Figure 4.2 shows, this will involve an extremely ambitious jump in the level of access for a measure that has conventionally advanced in a very slow linear fashion. This represents an important policy shift in rural drinking water supply, moving from an emphasis on expanding access, usually through handpumps, to an approach based on improving service levels, through piped schemes. Change at this pace and scale poses significant challenges to the viability of the community management model for rural drinking water supply in India.

Yet data at the national level in India obscures some significant differences from state to state. In Bihar state just 3 per cent of the rural population have access to a household connection while in the state of Sikkim this rises to over 80 per cent (Census of India, 2011b). This section therefore presents the results from an analysis of state–wise secondary data. It focuses on two key dependent variables which are the improved water supply coverage rate and the household connection coverage rate, both as reported by the Census of India (ibid.). These are then assessed by how they relate to other development variables which are outlined in Table 4.1. The analysis used data that was available at the state level in India either through the academic literature or published by reputable institutions such as the Government of India or international bodies, such as the UN. These were then analysed using the Kendall's tau test with the results presented in Table 4.1 and explained below.

The analysis showed that there is no significant correlation between states with improved water supply and those with high household piped water supply coverage, suggesting that there are different drivers for each of these factors. For household piped water supply, there are four states above 75 per cent coverage, eight states in the 50–75 per cent range, seven states at 25–50 per cent and the rest below 25 per cent. On this measure GDP per capita has a highly significant, strong effect suggesting it is only in states that have sufficient levels of wealth that the transition to household piped water supply is being achieved. The same relationship between wealth and improved water source is not found. There are two likely reasons why:

1   the lower level of capital and recurrent resource commitment and institutional capabilities that are required for basic improved services mean they are within reach of lower wealth levels;
2   due to the inherent human right and constitutional requirement for providing improved access, there is greater emphasis on transfers from the Federal Government, international donors and non-state actors like NGOs to support these types of services meaning domestic-state wealth becomes less important.

In comparison to piped water supply, there is limited variability in the improved water supply coverage rates with 21 out of the 29 states scoring between 90 and 100 per cent coverage rates. However, out of the eight states scoring below 90 per cent seven of them come from the mountainous regions of Northeast India or the Himalayas (with the only exception Rajasthan that includes large areas of desert). This indicates that India is making progress towards universal access to improved water supply across all states, including the very poorest, apart from in areas where the mountainous geography becomes a key factor. This was confirmed by the highly significant association found between the physiographic variable and improved coverage as reported in Table 4.1. Although there could be different reasons for this, it is speculated that there could again be two overarching reasons:

1   mountainous physiographic settings tend to not be conducive to the basic improved water source (handpumps) but rather served by more complex and expensive gravity-fed piped systems making the level of investment for improved access higher in these regions;
2   inaccessibility of some villages in mountainous physiographic settings further increases the challenge of delivering basic services as compared, on average, to other physiographic settings.

Table 4.1 Results of the bivariate analysis of development indicators against rural water supply coverage

Macro-development indicator	Household piped water supply			Improved water supply			Shapiro–Wilk (test for normality)
	Kendall tau	Significance	Interpretation	Kendall tau	Significance	Interpretation	
1. Household piped water supply	1.000	n/a	n/a	−0.023	0.865	Not significant	0.000
2. Improved water supply	−0.023	0.865	Not significant	1.000	n/a	n/a	0.039
3. GDP per capita (PPP)	0.622	0.000	Large effect (positive), highly significant	0.117	0.385	Not significant	0.012
4. Human Development Index	0.395	0.004	Medium effect (positive), highly significant	−0.277	0.049	Small effect (negative), significant	0.097
5. Devolution Index (rank)	−0.013	0.925	Not significant	−0.095	0.523	Not significant	0.637
6. Below poverty line (rural population)	−0.498	0.000	Medium effect (negative), highly significant	−0.031	0.821	Not significant	0.034
7. Literacy rate	0.364	0.006	Medium effect (positive), highly significant	−0.026	0.850	Not significant	0.806
8. Gini-coefficient	0.117	0.377	Not significant	0.387	0.004	Medium effect (positive), highly significant	0.086
9. Growth in poverty elasticity (2005–2012, rural below the poverty line, state GDP)	0.193	0.159	Not significant	−0.078	0.567	Not significant	0.000
10. Physiographic zones	0.130	0.401	Not significant	−0.631	0.000	Large effect (negative), highly significant	0.000

It is accepted that this descriptive statistical analysis of trends does not provide any evidence of causation, nor does it assess the compounding of one variable onto another, yet it is still considered to provide appropriate insights into the overarching pattern of rural water access across Indian states. It helps illustrate how India has largely achieved its goals around universal improved water supply coverage apart from in the mountainous states. Beyond these states, however, differences in performance in terms of household piped water supply can best be explained by the development status of the states. While no single variable can predict success, GDP per capita (PPP) and the 'below the poverty' line measures are most strongly associated with household coverage rates. It is argued in the next section that these factors are reflected in the political economy of states.

The political economy of development across Indian states

This section builds on the analysis above to argue that a key driver of differential performance in terms of development and rural water service coverage can be summarised as the political economy of the state. Using a political economy lens has become a common approach in a series of recent studies into rural water services (Chowns, 2014; Harris et al., 2011; Jones, 2015). This reflects the general trend in the sector to expand analysis beyond infrastructure and service levels, to understand the context in which rural water services are provided (Lockwood and Smits, 2011). Overall, this research develops an approach that links the empirical analysis above with studies on the political economy of India, in particular Kohli's (2012) empirically-driven 'State-Society Framework' on the differences between Indian states. In his work Kohli argues that Indian states can be allocated into three broad (political economy) categories that reflect the way in which 'development dynamics' emerge in each to drive development outcomes. The political economy categories are 'neo-patrimonial', 'social democratic' and 'developmental' and are considered to provide an empirically valid and analytically useful set of categories for analysing state-wise trends in India. Each is described below.

Neo-patrimonial states

In the 1980s the Indian demographer and economist, Ashisha Bose, coined the phrase 'BIMARU' to describe the states of Bihar, Madhya Pradesh, Rajasthan and Uttar Pradesh. As this acronym sounded like the Hindi word for 'sick' it became the popular shorthand in the press for underperforming states with poor governance records (Sharma, 2015). It later became expanded to 'BIMAROU' to include the state of Odisha and, since its popularisation, the states of Chhattisgarh and Jharkhand have been carved out of Bihar and Madhya Pradesh, respectively. It is these BIMAROU states that are considered to have the greatest 'neo-patrimonial' tendencies in India (Kohli, 2012). The term has most commonly been used in the broader development literature to describe the political economy of post-colonial Sub-Saharan African states (Englebert, 2000; Kohli, 2012). The concept is rooted in an understanding of traditional political authority which is based on a leader's ability to distribute resources to supporters in return for political support (Kelsall, 2011). This can be described as clientelism but Max Weber used the term 'patrimonialism' (Weber, 1978). It has since been reapplied as 'neo-patrimonialism' to describe a situation where this type of 'personal' political relations 'overlaps' with formal, impersonal forms of governance (Kelsall, 2011).

Kohli (2012) argues that neo-patrimonialism has taken particular root in these states of India because politics (and business) have been dominated by the high, often land-owning, castes who secure broader political support through short-term patrimonial strategies. In such a context corruption and clientelism become widespread, retarding the development process. Such an analysis is reinforced by looking at the developmental statuses of the BIMAROU states with lower levels of human development found in these states compared to broader India. Before proceeding it is important to acknowledge however that 'the problems of India's neo-patrimonial states need to be kept in perspective; India is no Congo or even a Zimbabwe' (Kohli, 2012, p. 154). Nationally, India is now ranking mid-way in the Corruption Perceptions Index (Transparency International, 2014) at 85 out of 175 countries. Yet it is still argued that neo-patrimonial characteristics are strong in the BIMAROU states of the country and that this political economy 'model' has shaped and is shaping developmental processes, including in rural water services.

Social-democratic states

In the states that exhibit social-democratic principles, political and economic power is dispersed more equally throughout the society. This is reflected in the ability of these states to be comparatively more effective at reducing poverty and promoting human development, even though the absolute level of economic wealth may be lower than some other states. The classic example is Kerala which is only the ninth richest as measured in GDP per capita yet has the highest HDI score and literacy rates. The 'social-democratic' mechanisms that underpin this come through the way in which almost all social classes and caste groups are effectively enfranchised within the democratic system (Kohli, 2012). As political elites generally have a broad base of support in these states it encourages them to favour policies that promote equitable development outcomes – while the generally high literacy rates and engagement in politics means the population is able to hold them to account (Desai, 2006). Overall, this means there is a well-established sense of 'public purpose' within these states that is reflected in a redistributive approach to development (Kohli, 2012).

While Kerala is the standout example of the social-democratic class, there are other examples, with Kohli (2012) arguing that all the South Indian states exhibit some social-democratic characteristics in that the political system is less likely to follow entrenched caste-based politics (as is the case in many Northern states, especially those in the Hindi heartlands such as Bihar and Uttar Pradesh). There are a number of other states that perform well in terms of having low levels of poverty to GDP ratios. These include Punjab, Andhra Pradesh and Himachel Pradesh. Yet, as argued by Kohli (ibid.), this type of analysis only moves towards an indicative classification as there are exceptions to the rules. For example, West Bengal is often considered a highly social-democratic state because lower classes and 'marginal' caste groups are well-represented in the political system. However, this has not translated well into strong development outcomes (Desai, 2006). This shows that the political economy classification should be treated as useful for explaining broad trends but that it is not a universal, all-encompassing explanation.

Developmental states

The final category of political economy classification of Indian states proposed by Kohli (2012) is the 'developmental' state. The use of the term developmental state first emerged to explain the remarkable rise of East Asian nations, such as Japan and South Korea, in the second half of the twentieth century. Johnson (1982) famously conceived the concept, arguing that the key to Japan's post World War II economic success was that the government took an interventionist approach to supporting capitalist production, as compared to the neoliberal – '*laissez-faire*' – economic policies that were promoted by Western countries or the 'socialist' centrally planned approach of Soviet states. This close alliance between government and business is in some ways found right across India as the extent of government bureaucracy means that the business class often has close alliances with government officials. Yet in some states these alliances are more strongly reflected in a strategic consensus between business and government with a collective aim to promote economic growth as the driver of development (rather than elites focusing on redistributive policies or serving piecemeal political and personal interests) (Kohli, 2012).

Gujarat and Maharashtra are the archetypal cases where there is particularly close cooperation between the business and political elites. These developmental models have led to significant public revenue generation through the tax system that in turn has enabled higher levels of investment in public infrastructure, especially the type of infrastructure that supports growth such as roads. Yet another key characteristic of these developmental states is that they enjoyed 'first mover' advantage in terms of economic growth as they represent the traditional heartlands of economic prosperity in India, so it is hard to assess whether these developmental policies could be translated to the poorer states that lack comparative advantage. Either way, the more top-down and elite-led approach to development is something that is thought to have significant implications for 'best-fit' thinking in terms of rural water services in developmental contexts.

Overview of the political economy of Indian states

In the section above three broad political economy models were described. However, it is perhaps more helpful to consider them as political economy 'tendencies' rather than standalone categories. In every state each of these tendencies exists to a greater or lesser extent so they should not be considered distinct or static categories. Political trends shift over time, especially in terms of the party politics of the states. Yet the broad pattern of political tendencies described previously has persisted for long enough to lead to consistent associations with developmental outcomes (Kohli, 2012). This is considered to be particularly relevant for the 'extreme' cases such as neo-patrimonial Bihar, social-democratic Kerala or developmental Maharashtra. The categories become more nuanced for states that do not exhibit very strong characteristics of a single category but more of a mixture of tendencies. Such an understanding of differences between Indian states, however, will be used to group the case studies presented in later chapters.

Clarifying the governance context for community management in India

The difference between the political economy of Indian states is partly reflected in the federal structure of governance in India. This section explains this in relation to rural water services. It first focuses on the governance of village regions before scaling-out to consider the state and federal systems. Contemporary trends in India's governance system can be traced back historically to the concept of *Swaraj* or 'self-rule'. Swaraj has been associated with rejecting the foreign rule of British and other colonial powers (Parel, 2011) but also has a related meaning in its application to village-level governance summarised in Gandhi's famous statement: 'My idea of the village Swaraj is that it is a complete republic independent of its neighbours for its own vital wants' (Mahatma Gandhi, as quoted in Bhatt, 1982, p. 87). This idea of autonomous village republics is reflected in the devolution agenda followed by the Government of India since the early 1990s (Johnson et al., 2005). This has involved devolving statutory powers to the Panchayat Raj Institutions (PRI) – the three-tier system of local self-government that was introduced in the previous chapter. As explained, the lowest level of the PRI is known as the Gram Panchayat[1] which operates at (or close to) village scale and is elected by all adult residents of the village (known as the Gram Sabha). Under this system, statutory responsibility is given to the Gram Panchayats for delivering public services including the provision of drinking water as well as 28 other areas such as street lighting (Government of India, 1993). For this purpose the Gram Panchayat usually sets up a number of sub-committees, such as a schooling and education committee. For rural water services it establishes a VWSC, which will assist and advise the Gram Panchayat, often on a voluntary basis, on the provision of these services (Government of India, 2012a). The key institutions at the village level are explained here:

- Gram Sabha: includes every person of voting age within a village. Usually, the Gram Sabha meets to take key decisions during the implementation of a water scheme and it is responsible for approving the plans that the Gram Panchayat and VWSC have for water supply each year.
- Gram Panchayat: the lowest level of government in rural India. It is part of the Panchayat Raj system of local self-government which promotes self-rule within Indian villages. Each Gram Panchayat has a president known as the Sarpanch who is elected by the members of the Gram Sabha. Typically, he or

she is supported by a vice president and clerk, while a number of elected ward members also sit within the main Gram Panchayat council. Together they are responsible for the provision of many public services within the village, including domestic water supply. The Gram Panchayat owns and manages (in partnership with the VWSC) the water supply with its tasks including: approving investment plans and getting financing; approving annual budgets and user fee charges after discussion in the Gram Sabha; approving contracts with operators; coordinating with the block and district support organisations; hiring trained mechanics for regular preventive maintenance for handpumps, and trained operators for piped water supplies (Government of India, 2012a).

• Village Water and Sanitation Committee (VWSC): a standing committee of the Gram Panchayat of between 6 and 12 members that takes on the responsibility for the everyday operation, maintenance and administration of the water supply service. It is chaired by the president of the Gram Panchayat and includes some ward members – it should also include at least 50 per cent women and representatives from all social classes and castes with the village. The existing members nominate new members onto the committee but any decision must take into account the predetermined quota system. Key tasks include: collecting household contributions and user fees; opening and managing a bank account; preparing annual budgets and recommendations for user fee charges; organising people to be vigilant about not wasting water and keeping water clean; ensuring professional support for handpump caretakers and piped water supply operators; ensuring access to spare parts for handpumps and trained mechanics for regular preventive maintenance; ensuring the operators handling piped water supply systems are provided with adequate training to gain the technical and financial skills needed to do the job (Government of India, 2012a).

The Gram Panchayat is considered to represent a form of local self-government rather than simply local government. The difference being that local self-government, which has its own elected officials and statutory powers over a considerable range of functions, has greater autonomy, compared to local government and more direct accountability to the local population (Datta, 2007; Rajesh and Thomas, 2012). The local self-government system had some historical precedent in pre-colonial and colonial times but its power had been eroded in much of the post-colonial period as the Government of India and state governments sought to drive a relatively centralised programme for rural development (Banerjee, 2013). This was until 1993 and the 73rd amendment to the Constitution of India reversed those decades of centralisation.

Since then India has advanced a strong devolution agenda so it is useful to reflect on the maturity and character of that process. Devolution is a strong form of decentralisation that involves the formal statutory transfer of powers to lower levels of government (James, 2011; Robinson, 2007). However, the transfer of these functions should also be accompanied by the appropriate transfer of funds to finance the activities and functionaries to undertake any necessary work. Following 25 years of devolution there are concerns that the devolution of power to Panchayats has been hampered by a lack of devolution of funds and functionaries (Banerjee, 2013). Such concerns are not universal, with key differences between the states. Some states, such as social-democratic Kerala, are renowned for advanced and genuine devolution to the Panchayat Raj Institutions (Heller et al., 2007), whereas other states retain a largely centralised character in public administration. Each year the Ministry of the Panchayati Raj publishes a devolution list that ranks states against one another depending on the devolution of functions, functionaries and funds. It shows there is no pattern in terms of the level of devolution and the wealth of a state. This is one of the key challenges of conducting research across Indian states – although there is a federal government the set-up for governance within each state can vary considerably. Part of these differences relate to how the system of local self-government functions alongside the centralised agencies of the state government, which are now explained below.

At the federal level there is a Ministry of Drinking Water and Sanitation that sets policy, budgets and oversees the entire drinking water and sanitation sector (James, 2011). The administration of the country is then divided into 29 states, with nine of these having populations of over 50 million people. A cabinet of state ministers lead departments in the various domains of public administration, and would usually include a State Rural Water Supply Agency. These agencies set the budgets and overall plan for rural water supply in the state. They are meant to provide technical services to the Gram Panchayat but often retain decision-making power over budgets and technical design (Banerjee, 2013). This is considered to be partly a hangover from the supply-driven era of expansion of rural water services prior to the devolution of the 1990s. Through research into the functioning of such agencies within the sanitation sector, Hueso and Bell (2013, p. 1013) described these organisations as representing the 'technocratic governing machinery' of the Indian state. They argued that they represent 'a hierarchical and technocratic bureaucracy that is well suited to send down technical designs and subsidies for physical infrastructure projects' (ibid.). In this sense, the paradox in the governance of rural water services in India is the extremely advanced devolution of local self-government alongside centralised and often bureaucratic State Rural Water Supply Agencies. The balance between these decentralised and centralised agencies is specific to each state and shapes how rural water services are governed within them.

Chapter summary

This chapter has focused the book on the Indian context. It has explained that India is a vast and populous country that has varying levels of human development and access to rural water services. The diversity means the research can investigate community management across different contexts. To provide explanatory insights into why such differences can exist in a single country, this chapter has provided an overview of the political economy tendencies that can be found in India. It was argued these represent a useful contextual background when considering trends across case studies from different states. The governance system for rural water services was also explained with the tension between a highly decentralised mode of local self-government and the legacy of centralised State Rural Water Supply Agencies highlighted.

Acknowledgements

Part of this chapter augments another research paper published by the authors: P. Hutchings, R. Franceys, S. Mekala, S. Smits and V. James (2017) 'Revisiting the history, concepts and typologies of community management for rural drinking water supply in India', *International Journal of Water Resources Development*, 33(1), 152–169.

Note

1   Although there is some variety in practice, such as in the Tribal belts in the North Eastern States where the Sixth Schedule to the Constitution of India means that village administration can follow a different model involving the formation of a tribal council known as a Durbars (Constitution of India, 1950).

5 Case study research methodology

The full case study methodology for the Community Water Plus project is presented as a working paper (Smits et al., 2015) that is augmented into a series of fieldwork protocols. This chapter builds on those documents to provide an overview of the research methodology. Based on the research context and conceptual understanding outlined in the previous chapters, this research seeks to obtain insight into the type, extent, style and costs of supporting organisations that are required to ensure sustainable community-managed water service delivery. It therefore, by definition, focuses on 'successful' cases of community management and support of rural water supplies, in order to be able to assess what support was provided and with what resource implications. We focused our research on the resource implications at the 'enabling support entity' level (recognising that there may not always be a clear distinction between the support and service delivery levels) and also recognise that we have to be sure that we are indeed investigating successful community-managed service delivery. What can be considered successful can be understood at various levels: at the level of service that users receive, at the level of the service provider carrying out its tasks with a certain degree of community engagement, and at the level of the support agent in partnership with the service provider. In order to answer the research questions, the research therefore assesses the degrees of success of various elements, as summarised in Figure 5.1, recognising that the key validation of success is the service achieved by consumers.

The research took place at two key levels of analysis:

1   the programme or support model; and
2   the community level.

The research seeks to analyse examples of community management *plus* that are successful at scale. We realise that there are always individual communities that – with or without support – may achieve sustainable services. However, what interests us are examples where community management *plus* has led to sustainable services at scale, that is where a significant number of communities are being successful according to the criteria identified in Chapter 2 related to effectiveness, sustainability and replicability. The implication is that our first unit of analysis is the area served by a particular programme and/or support model. We recognise that this

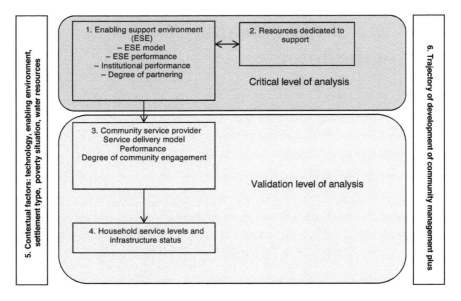

Figure 5.1 Elements of the research

support typically comes in different forms from government (national, state and local-self), NGOs, private actors and hybrids. It is to be expected that the degree of success in any model depends to a large extent on the resources dedicated to these functions. This refers both to the monetary costs (as per the cost categories) as well as non-monetary costs, such as the presence of skilled staff and political capital. We quantify the financial resources and provide a qualitative description of other resources, such as human and political capital, that are spent on this.

However, the other assessments (i.e. the ones related to the service provider, degree of community engagement, service levels and contextual factors) need to take place at the second unit of analysis, the community level, that being the level where data on service providers and service levels are generated, and where contextual factors are best captured. We recognise that villages are not always clearly defined. Within a village there may be sub-communities, or the boundaries between two villages may not be very clear. For purposes of this research, the defi- nition of village as per the census is followed. A village may contain one or more water systems – particularly, the combination of communal piped supplies with individual point sources is common. Taking the village as a unit of analysis thus means assessing the aggregate level of service from these various systems, as well as the aggregate performance of all service providers in that village.

By definition, a service level refers to what users actually receive. Therefore, data on service levels in part needed to be collected from a third level: the household. From there, the service level is constructed from the bottom up, i.e. data from households is aggregated to community level. A particular concern is equity in service delivery, among others how different groups within a community have access to a service and how they participate in the management of the service. This also requires compiling data from households, so as to see the variability in service levels and participation within a village. Again, this requires data collection at household level, so as to analyse variation and difference between households at village level. This means that the household is not a third unit of analysis, but a level of data collection.

We recognise that what may be required to be successful in one case may not be adequate to be successful in another. Context matters – for example, the management of a more complex multi-village scheme may require a higher degree of professionalisation and support than a simpler handpump system. Likewise, with similar inputs differing degrees of success may be achieved in different types of settlements: for example, with a relatively small amount of external support, success may be achieved in a relatively well-off settlement where people are willing to pay for water and where there are no water security risks. In a village with a large desti-tute population or one with serious water resources management issues, a similar small amount may not be sufficient to achieve success – or maybe it will, if that community is well organised. In order to understand the type and extent of support that is needed to achieve successful service delivery, one needs to relate them to these kinds of contextual factors. In our research, we assess these factors, including among others: type of technology employed, the socio-economic and poverty status of the community, the spatial dimensions of the type of settlement and the water resources situation.

Case selection and sampling

Reflecting the levels of analysis given above, the approach to sampling involved a two-tiered selection process. The first layer was to select programmes from across India, with the second level identifying the villages to study within these programmes. In sampling the programmes for this study, we aimed to select programmes that deliver best-practice community management plus across various social, technical and geographical spectrums. The case selection process led to the selection of 20 case studies that are listed in Table 5.1. The selection for that final list were made from an all India sampling frame of 92 cases that was developed from the literature and feedback from government and other sector stakeholders working in the Indian sector (and which formed the basis for the review of Indian community management presented in Chapter 3). The final case studies were then selected through consultation with local officials and pilot visits were made by individual research teams to verify that they would make an appropriate case study. An initial condition of selection was that the cases should have been operational for at least five years. In the end, the average age of case study was seven years but there were five that had been operational less than five years.

In selecting the cases a key stratification was to cover different GDP per capita measures at the state level with 17 different states being covered. These ranged

Table 5.1 List of Community Water Plus case studies

State	Case name and reference	Technology type	Household connection in case villages (% from survey)	Years of operations
1. Jharkhand	Drinking Water and Sanitation Department (Ranchi West Division), Ranchi West District, Jharkhand (Javorszky et al., 2015)	Single-village scheme (tubewell)	48%	4
2. Madhya Pradesh	VASUDHA and PHED support in Dhar district, Madhya Pradesh (Ramamohan Roa and Raviprakash, 2016a)	Single-village scheme (tubewell)	100%	2
3. Odisha	Gram Vikas model in Ganjam, Bargarh and Jharsuguda districts, Odisha (Javorszky et al., 2016)	Single-village scheme (tubewell)	95%	3
4. Chhattisgarh	Chhattisgarh Public Health Engineering Department, Rajnandgaon district (Javorszky, et al., 2015)	Single-village scheme (tubewell)	54%	14
5. Meghalaya	The Dorbars and gravity-based piped water supply in Meghalaya (Saraswathy, 2016b)	Single-village scheme (gravity-fed)	31%	9
6. Rajasthan	Swajaldhara programme in Jaipur district, Rajasthan (Harris et al., 2016a)	Single-village scheme (tubewell)	91%	6
7. West Bengal	Community-managed handpumps in Patharpratima, West Bengal (Smits and Mekala, 2015)	Tubewell handpump	0%	6
8. Telangana	Decentralised drinking water service delivery – community-managed water purification units in Telangana (Chary Vedala et al., 2016a)	Water kiosk	39%	5
9. Karnataka	Jal Nirmal and beyond: supporting the community management of rural water supply in Belagavi district, Karnataka (World Bank) (Ramamohan Roa and Raviprakash, 2016b)	Single-village scheme (surface)	65%	6

Table 5.1 continued

State	Case name and reference	Technology type	Household connection in case villages (% from survey)	Years of operations
10. Himachal Pradesh	Community Water Classic: the success of community-managed water supplies in Himachal Pradesh with limited on-going support (Harris et al., 2016c)	Single-village scheme (gravity-fed)	92%	6
11. Punjab	24/7 water supply in Punjab: international funding for local action (World Bank) (Harris et al., 2016b)	Single-village scheme (tubewell)	100%	3
12. Uttarakhand	Support to community-managed rural water supplies in the Uttarakhand Himalayas – the Himmotthan Water Supply and Sanitation initiative (Smits et al., 2016)	Single-village Scheme (gravity-fed)	0%	8
13. Kerala I – World Bank	Jalanidhi programme in Nenmeni Panchayath, Wayanad district, Kerala (World Bank) (Saraswathy, 2016a)	Multi-village scheme	100%	9
14. Kerala II – Local self-government	Community-managed rural water supply in Malappuram district, Kerala (Chary Vedala et al., 2016b)	Single-village scheme (surface)	100%	9
15. Gujarat – WASMO Gandhinagar	WASMO in Gandhinagar district (Chary Vedala et al., 2015a)	Single-village scheme (mixed)	100%	11
16. Gujarat – WASMO Kutch	WASMO in the desert Kutch region (Chary Vedala et al., 2015a)	Single-village scheme (mixed)	99%	6
17. Tamil Nadu – Local self-government	TWAD Board and the Panchayat Raj Institutions in Erode district (Saraswathy, 2015)	Single-village scheme (mixed)	93%	10
18. Tamil Nadu – Public-Private Hybrid	TWAD Board and the Hogenakkal Water Supply and Fluorosis Mitigation Project in Morappur district (Hutchings, 2015)	Multi-village scheme	94%	3

Table 5.1 continued

State	Case name and reference	Technology type	Household connection in case villages (% from survey)	Years of operations
19. Maharashtra	Maharashtra Jeevan Pradhikaran (MJP) and the Shahnoor Dam project (Chary Vedala et al., 2016c)	Multi-village scheme	100%	14
20. Sikkim	Decentralised local self-government and gravity-based piped water supply in Sikkim (Saraswathy, 2016a)	Single-village scheme (gravity-fed)	0%	5

from the poorest state in the study, Jharkhand, having a 2014 GDP per capita of $2,632, to the richest state Sikkim, at over $10,000 GDP per capita. These are comparable to levels found in Chad and Albania (World Bank, 2016), respectively, showing the study covers a range of wealth from a poor Sub-Saharan Africa context to a middle-income Eastern European country. Selecting by state also ensured the study had a good geographical coverage of different regions in India, as shown in the map in Figure 5.2.

Beyond wealth, two other selection stratifications were to cover different types of support programmes, from government support programmes to small-scale NGO programmes. The full range of different support systems is the subject for discussion in the following chapter so will not be explained here. The stratification for the technical design of the system was designed to cover different forms of piped supply and also non-piped supply. In considering how to design the technical stratification criteria there was a consideration of adopting the WASHCost categories (McIntyre et al., 2014) yet these were not widely recognised in India so it became easier to adopt the Indian terminology for different types of water supply set-ups. These were:

* *Borehole handpump* – a borehole with handpump attached.
* *Single-village scheme (tubewell)* – reticulated water supply serving the population of a village through piped water supply either to household connections or public stand-posts (or both). The source for the system is a local borehole.
* *Single-village scheme (surface)* – reticulated water supply serving the population of a village through piped water supply either to household connections or public stand-posts (or both). The source for the system is a surface water source usually a river in the plains.
* *Single-village scheme (gravity-fed)* – reticulated water supply serving the population of a village through piped water supply either to household connections or public stand-posts (or both). The source for the system is a spring or similar in a hilly or mountainous area.

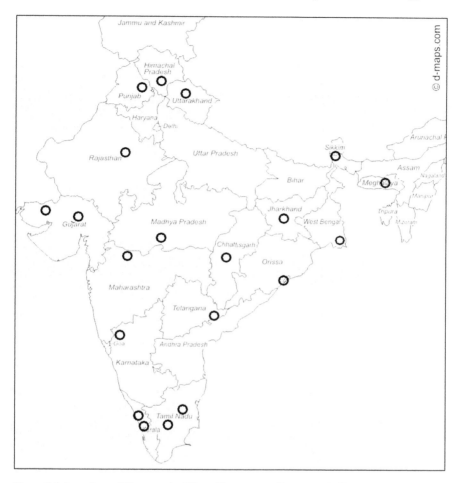

Figure 5.2 Location of Community Water Plus case studies across India

- *Single-village scheme (mixed)* – reticulated water supply serving the population of a village through piped water supply either to household connections or public stand-posts (or both). The source for the system comes from a mixture of borehole, surface and/or gravity-fed sources.
- *Multi-village scheme* – reticulated water supply serving the population of multiple villages through piped water supply either to household connections or public stand-posts (or both). The source for the system can come from either borehole, surface or gravity-fed sources.
- *Water kiosk* – a type of point source in which users pay to access water from a vending point in the village as they use it. The source for the system can vary but usually a Reverse Osmosis plant or equivalent treats the water before vending.

A problem in specifying the technical design of a water system in India is that often multiple systems can overlap and these can even be managed by different entities (Ratna Reddy and Batchelor, 2012). This research focused on the technical system that was implemented and/or maintained under the enabling support environment being studied. As shown in Table 5.1, the study covers different forms of piped supply adequately but is weak on other forms of supply, such as handpumps and water kiosks, with only one case study each focusing on these, respectively.

Within each case study there was an embedded case study design (Yin, 2003) in which a number of community service providers which are supported by the enabling support environment were studied. The intended design was for four such community service providers to be covered across three 'programme villages' and one 'control village' coming from outside the programme, although, there is some variation on this arrangement across the cases. Based on the selection of the community service providers there was intended to be 80 villages (60 programme and 20 control) as part of the study and so there was a need to collect data from households in the programme and control villages so as to validate the service levels people received. For this purpose a predefined sample size of 30 households was selected for each village, which was shaped by the resources available to the project and the desire to achieve a reasonably representative sample (which will be discussed below). Random interval sampling was followed based on standard practice for such household survey research in developing countries (Deaton, 1997).

Data collection, processing and analysis

The data collection methods for fieldwork are outlined in the fieldwork protocols which explain that the following methods were used at each analytical level:

- *Enabling support environment* – key informant interviews and secondary records (i.e. accounts, reports).
- *Community service providers* – key informant interviews, focus groups and secondary records (i.e. accounts, reports).
- *Household level* – household surveys.

On average in each case study 14 key informant interviews, 7 focus groups and 118 household surveys were conducted by the field research teams. This gave a total dataset of 280 interviews, 140 focus groups and 2239 household services collected across 20 case studies.

The data collected in each case study was processed by the researchers through a series of data-processing tools and databases to provide a framework for comparison. To measure the professionalisation of the enabling support environment and the community service provider, the research adopted an approach called qualitative information systems (Postma et al., 2004; da Silva Wells et al., 2013). These were developed as a means for standardising evaluation in water, sanitation and hygiene programmes. The researchers were provided with a series of questions to ask in key informant interview and focus groups, and then in the analysis of the data collected they are asked to incorporate a quantitative logic in their assessment of the findings. In essence, they were asked to complete a ranking ladder, similar to a Likert scale. Participation and partnering were integrated into data-processing tools based on laddered frameworks that were used to structure ranking systems for the researchers. Both these frameworks are given in the appendix. The other similar data-processing tool was for the enabling support environment. It related to desire to deepen the characterisation of successful enabling support entities by assessing the relative strengths and weaknesses of them using an institutional assessment (Cullivan et al., 1988). Such a tool was originally developed by USAID to assess the performance of urban utilities but it was considered potentially useful to adapt to this new context.

The research also analysed service levels and financial data within and across case studies. The calculation of the service levels was based on using household survey data to calculate the service level benchmarks presented in the previous chapter that contains five different parameters: quantity, accessibility, quality, reliability and continuity. Amin et al. (2015) proposed a composite service level indicator that brings these different parameters together into one measurement unit. To avoid over compensability between the different parameters (i.e. high reliability making up for low quantity), the approach taken draws on a mixture of nominal threshold logic and the geometric mean. This approach means that if any service level is either substandard or no service then the composite indicator cannot exceed this reflecting a nominal logic. However, above this, the geometric mean is employed as it 'limits the impact of compensability and eclipsing (where the composite indicator is insensitive to a single variable)' (Amin et al., 2015, p. 14). This approach is used as the main benchmarking measure of service levels when comparing case studies.

The final element of the synthesis framework relates to standardising the assessment of financial costs. Researchers were asked to collate the data related to the various elements of the 'life-cycle cost approach' described in Chapter 2. The focus was on financial costs only rather than economic costs was designed to ensure proximity to the actual data rather than have individual researchers attempt to calculate items such as the opportunity-cost of voluntary labour. The costs were asked to be produced as follows: CapEx – per person, in 2014 Indian rupees, and recurrent costs (OpEx, OpExES, CapManEx) – per person, per year in 2014 India rupees. The 'per person' scale provided a base for comparison but the data has been converted to US dollar prices for the purpose of this book. Analysis of that data and other data followed basic descriptive statistical testing to present basic frequencies and central tendency measures alongside measures of dispersion. Where appropriate, these basic descriptive statistics were accompanied by non-parametric statistical testing of the differences between groups and bivariate correlations. The specific tests used are presented in Table 5.2.

Chapter summary

This chapter has provided an overview of the research methodology. In doing so, it has introduced the case studies that were analysed as part of this research. For the

Table 5.2 Data analysis approach summary

Category	Enabling support environment and community service provider	Service-level analysis	Financial costs	Overview analysis
Data source	20 – databases consolidated into overview database	2,239 household surveys	20 – databases consolidated into overview database	Overview database
Critical data type	Qualitative-quantitative (ordinal or categorical)	Ordinal (composite service level indicator)	Continuous	Continuous v. ordinal (treated as continuous)
Basic frequency	Count and per cent	Count and per cent	Count and per cent	Count and per cent
Central tendency	Mode or median	Median	Mean	Mean
Dispersion	Range and inter-quartile range	Standard deviation, inter-quartile range	Standard deviation, inter-quartile range	Standard deviation
Shape of distribution	N/A	Visualisation, skewness and kurtosis	Visualisation, skewness and kurtosis	Standard deviation, inter-quartile range
Statistical tests for comparing average of groups within one variable	N/A	Mann–Whitney U-test or Kruskal–Wallis test	Mann–Whitney U-test or Kruskal–Wallis test (with post hoc Dunn-Bonferroni pairwise testing)	Mann–Whitney U-test or Kruskal–Wallis test (with post hoc Dunn-Bonferroni pairwise testing)
Statistical tests for association between two variables	N/A	Kendall's tau	Kendall's tau	Kendall's tau
Level of significance	N/A	<0.05 significant and <0.01 very significant	<0.05 significant and <0.01 very significant	<0.05 significant and <0.01 very significant

presentation of the initial case studies, the book has adopted the political economy framework to help organise the cases into groups for initial analysis. These include the neo-patrimonial, social-democratic and developmental states – and also a chapter on the mountain and hilly regions case studies. The following chapters tackle each set of cases in turn.

Part II

Community management case studies of success from India

6 Community management in the 'neo-patrimonial', low-income states

This chapter considers the community management case studies from the 'BIMAROU' states which make up the economically poorest states in this study. They include Jharkhand, Madhya Pradesh, Odisha, Chhattisgarh and Rajasthan. The support programmes detailed in this chapter include the large public agencies responsible for rural water supply that are usually called public health engineering departments (PHEDs) but can also be called variants on this name. In Jharkhand, Chhattisgarh and Rajasthan these are the support entities assessed and as they are state-wide programmes they also serve the control villages that have been assessed. There are also two cases where support is provided through NGOs, either in partnership with the PHED as in Madhya Pradesh, or with an NGO becoming the government-approved support entity, as is the case in Odisha. In both cases the NGO-involved programmes are significantly smaller in scale than the PHED programmes (see Table 6.1).

Government-supported community management in Jharkhand, Chhattisgarh and Rajasthan

The government-supported programmes in this chapter help illustrate the differences between two of the recent Government of India flagship rural water supply programmes. The Rajasthan case provides insight into the earlier Swajaldhara (2002–2009) community management programme that followed the 'demand-responsive approach' with communities expected to contribute 10 per cent of capital costs and operate schemes in relative independence (Government of India, 2003). In contrast, the Jharkhand and Chhattisgarh cases showcase the National Rural Drinking Water Programme (NRDWP) (2009–present) that has shifted back to a more recognised supply-driven model with a stronger role for the GP (Government of India, 2013a). Figure 6.1 helps illustrate the differences between the two with a simplified schematic of the organisational set-up across the programmes. As it shows, the arrangements are similar in that they both have a PHED as the primary support entity for community management but in the NRDWP there is also meant to be a new organisation labelled a 'block resource centre' that can provide specialist capacity building and broader software support to villages. However, in both the NRDWP cases presented in this chapter, these

Table 6.1 Overview of the enabling support environment and community service providers

Case no. (inverse GDP rank)	Case name	State	Enabling support environment	Community service provider	Scale of support programme (approximate population)
1	Drinking Water and Sanitation Department (Ranchi West Division), Ranchi West district, Jharkhand	Jharkhand	Drinking Water and Sanitation Department and Government (of Jharkhand)	Village water and sanitation committees	25,000,000
2	VASUDHA and PHED support in Dhar district, Madhya Pradesh	Madhya Pradesh	Vasudha Vikan Sansthan (NGO) and PHED (Government of Madhya Pradesh)	Drinking water sub-committee	20,000
3	Gram Vikas model in Ganjam, Bargarh and Jharsuguda districts, Odisha	Odisha	Gram Vikas (NGO, Odisha)	Village water and sanitation committees	350,000
4	Chhattisgarh Public Health Engineering Department, Rajnandgaon district	Chhattisgarh	PHED (Government of Chhattisgarh)	Gram Panchayat	35,000,000
6	Swajaldhara programme in Jaipur district, Rajasthan	Rajasthan	PHED (Government of Rajasthan)	Village water and sanitation committees	50,000,000

organisations were not found to be providing any direct support in the studied villages.

The Rajasthan case study represents an early example of the Swajaldhara programme in Jaipur district. The programme had the following prescribed nationwide guidelines (Government of India, 2003, p. 3):

(i)   adoption of a demand-responsive, adaptable approach along with community participation based on empowerment of villagers to ensure their full partici-

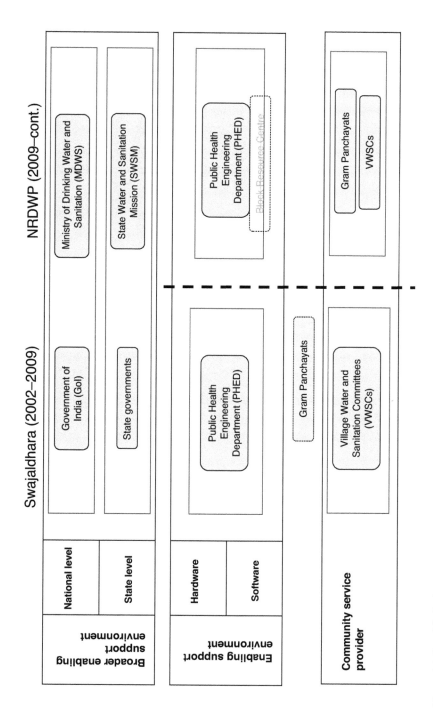

Figure 6.1 Simplified schematic of institutional arrangements for government-supported programmes in the neo-patrimonial cases

pation in the project through a decision-making role in the choice of the drinking water scheme, planning, design, implementation, control of finances and management arrangements;

(ii)   full ownership of drinking water assets within appropriate levels of Panchayats;

(iii) Panchayats/communities to have the powers to plan, implement, operate, maintain and manage all Water Supply and Sanitation schemes;

(iv)   partial capital cost sharing either in cash or kind including labour or both, 100 per cent responsibility of operation and maintenance (O&M) by the users;

(v)   an integrated service delivery mechanism;

(vi)   taking up of conservation measures through rain water harvesting and ground water recharge systems for sustained drinking water supply; and

(vii) shifting the role of Government from direct service delivery to that of planning, policy formulation, monitoring and evaluation, and partial financial support.

Notwithstanding the nationwide guidelines, the implementation of the programme remained a state government responsibility leading to variability in its implementation across the country (James, 2004, 2011). In Rajasthan the PHED took responsibility for Swajaldhara following its pre-existing staffing structures that remained largely technical and engineering-driven (Harris et al., 2016a). Through the programme it effectively became a support agency for the implementation period of a new project but played a very limited role in on-going support to the VWSCs who were responsible for operation and maintenance. In the villages studied, the Swajaldhara programme led to the construction of new borehole-fed single village schemes (SVSs) with communities contributing 10 per cent of the capital costs (ibid.). Following implementation the infrastructure assets were handed over to VWSCs for operation and maintenance with the expectation that the community would cover 100 per cent of the operational costs through tariffs. The systems have now been operational for the over ten years and, in this sense, the case demonstrates that relatively 'independent' community management can work to some extent.

Yet the VWSCs are now facing significant challenges that threaten the on-going viability of the services. As the district is close in proximity to the growing city of Jaipur, the Rajasthan state capital, the villages are becoming more urbanised (and richer) and this is driving an increase in demand for water at the household level (Harris et al., 2016a). Community members are increasingly turning to alternative sources, such as private boreholes or tankers, which threatens the financial sustainability of the programme through reduced tariff collection. Coupled with, and confounding, this financial threat the broader problem of groundwater source depletion is limiting the yield of the boreholes that supply the systems. As the Swajaldhara programme has finished, the PHED considers these villages to be working through the 'old system' so is providing limited support in addressing these challenges. Instead, it is now focusing on expanding bulk water provision from surface water sources across the district with the aim to shift villages from single village schemes (SVSs) to multi-village schemes (MVSs) over the coming decades (Harris et al., 2016a). In this sense, the case provides insight into communities that have managed piped water supply for over a decade but are now facing the challenges of system augmentation, as well as asset renewal, in an institutional context that has changed considerably since the initial implementation period.

In Jharkhand and Chhattisgarh the case studies reflect the support provided through the NRDWP (Government of India, 2013a, pp. 1–2) that sets out how the government must:

(i)   enable all households to have access to and use safe and adequate drinking water and within reasonable distance;
(ii)   enable communities to monitor and keep surveillance on their drinking water sources;
(iii) ensure potability, reliability, sustainability, convenience, equity and consumers' preference to be the guiding principles while planning for a community-based water supply system;
(iv)   provide drinking water facility, especially piped water supply, to Gram Panchayats that have achieved open defecation free status on priority basis;
(v)   ensure all government schools and anganwadis have access to safe drinking water;
(vi)   provide enabling support and environment for Panchayat Raj Institutions and local communities to manage their own drinking water sources and systems in their villages; and
(vii) provide access to information through an online reporting mechanism, with information placed in public domain to bring in transparency and informed decision-making.

In a similar vein to the Swajaldhara programme, although the NRDWP is a national policy, responsibility for its implementation is with the state government and in both Jharkhand and Chhattisgarh the PHED is the main government body charged with delivering it.

In Jharkhand the PHED has the twin objectives of implementing new rural water supply schemes and supporting communities to take on service provision as per the NRDWP guidelines. The Jharkhand case study comes from Ranchi district which is close to the state capital but which was selected as the area saw the earliest examples of where the NRDWP was implemented from 2010 onwards. In the programme villages, pre-existing borehole-fed SVSs that were constructed at no cost to the community were handed over to the local self-government institution, which then established a sub-committee, namely the VWSC, to take on the role of service provider. Like Rajasthan the PHED has largely retained its technical staffing structures but has tried to promote successful community service delivery through two interesting features that include smart institutional design of VWSCs and the use of 'incentivised subsidies'. The PHED stipulated that upon establishment of the VWSCs a *jal sahiya* (water volunteer) must be nominated from among the daughters-in-law of the village.[1] These women become treasurers of the VWSC and receive specialist training for this purpose. The general principle is that this approach minimises the turnover of a key administrative post, something that has been highlighted as a failure point for community management (Schouten and Moriarty, 2003). Subsidies were also used to encourage higher tariff collection rates in the first year of operation. The PHED provided a 'matched-grant' for the total amount of tariffs collected in that year and more broadly the VWSC continues to receive a 100 per cent energy subsidy as the electricity bill for water supply is paid directly by the PHED (Javorszky et al., 2015). In this sense, under the NRDWP in Jharkhand, communities have not been expected to pay 100 per cent of operational expenses but are still expected to play an active role in the operation of the system.

In the neighbouring state of Chhattisgarh a similar set-up of the enabling support environment can be found. However, here, there is a more advanced devolution of the Panchayat Raj Institutions that has shaped how support is provided and service provision conducted. This has led to a situation whereby the PHED takes the role of being primarily a construction agency that then hands over the infrastructure directly to the GP. Although the GPs are mandated under the NRDWP to establish an autonomous VWSC these were not established in the studied villages and so the GP takes on the role of service provider. The main method of capacity building is through a transition period in which the PHED operate the schemes for three to six months. There is no accredited training but the GP is expected to have appointed an operator who can shadow the PHED staff during this period. Similarly to Jharkhand, a series of subsidies are channelled toward service provision that mean, although the community contribute some level of tariffs, a major part of on-going support comes from government. In both of the Jharkhand and Chhattisgarh case studies the PHED provides periodic monitoring of water quality and functionality.

The government-supported programmes across these three case studies show that large engineering-focused public support entities have been charged with supporting community management in these states. The support provided has been implementation focused but this still has led to reasonable levels of success in that communities have taken on the management of service provision for periods between five to ten years.

NGO-supported community management in Odisha and Madhya Pradesh

The NGO-supported programmes for community management in the neo-patrimonial states come from the Madhya Pradesh and Odisha case studies. These programmes cover populations in the thousands rather than millions. In Madhya Pradesh the case covers an example of community management in which a state-level NGO called Vasudha Vikas Sansthan supplements the government PHED programme supporting communities to overcome fluoride contamination in local groundwater sources (Ramamohan Rao and Raviprakash, 2016a). In its work, rejuvenated open wells are used as an alternative source to deeper boreholes and connected to SVS distribution systems. In the Odisha case, Gram Vikas, an international flagship NGO of India, has been granted Project Implementing Agency status by the Government of Odisha and so operates as an agency delivering the government's water programme in certain areas of the state – particularly, those with high levels of tribal populations (Javorszky et al., 2015). Figure 6.2 provides a simplified organogram for each case.

Vasudha Vikas Sansthan has been operating in Dhar district, Madhya Pradesh, for the past ten years. Its water supply interventions concentrate on providing support to communities facing fluoride contamination in local groundwater sources. It has provided direct support to 20 villages where it has helped develop new infrastructure and software support to a further 35 villages to raise awareness around the dangers of fluoride (Ramamohan Rao and Raviprakash, 2016b). In the direct support villages the NGO has developed a work programme for reviving open dug wells as alternative sources for SVSs as these, generally, have lower levels of fluoride than the deeper groundwater sources. A community management approach is then followed for operation and maintenance with the NGO supporting drinking water sub-committees at the habitation level. Yet this support provided by Vasudha Vikas Sansthan must be considered as a Public-NGO Partnership as it is complimentary to the broader support provided by the PHED that continues to operate in the villages through the development and maintenance of handpumps (which cover 95 per cent of the rural population across the state; ibid.). In this sense, the villages studied have community-managed SVSs supplied by the revived wells as well as PHED-managed handpumps with some of these still providing potable water. It demonstrates an example whereby an NGO can provide supplementary support alongside government to address a specific need such as developing alternative sources when faced with groundwater contamination.

The Gram Vikas case study provides a rare example of an independent NGO-supported programme for community management that has reached some 'scale' in India – with the NGO having provided water and sanitation services to 350,000 people since 1992 (Javorszky et al., 2015). Originally established as a disaster relief organisation in 1979, Gram Vikas has been transformed into a flagship development NGO with a portfolio of development interventions. Through the 1990s and 2000s its work on water and sanitation was driven by the Government of India policy shift towards community management, particularly through the Swajaldhara programme, with Gram Vikas become a facilitating agency for Swajaldhara (ibid.). Since 2009 it has had state-government Project Implementing Agency status so it can continue to be funded through government programmes. However, as an NGO the organisation retains a greater autonomy to decide the approach it follows compared to a PHED. This is particularly telling in its intervention methodology known as MANTRA: 'Movement and Action Network for Transformation in Rural Areas'. This front-loaded approach is contingent on 100 per cent participation of communities in that prior agreement has to be made by every member of the community they support and will contribute toward the project before it commences. This means there is a high commitment threshold and the NGO retains the right not to

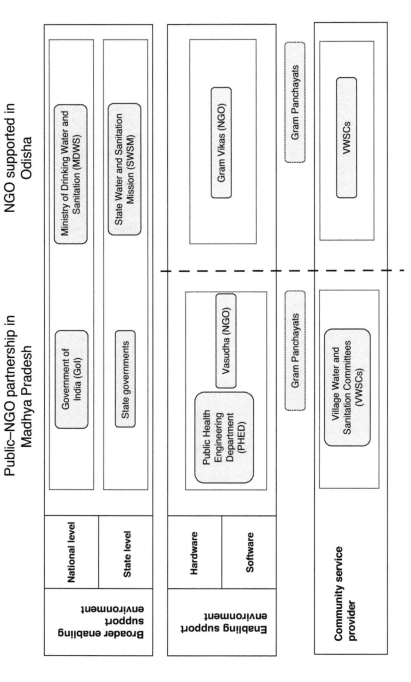

Figure 6.2 Simplified schematic of institutional arrangements for NGO-supported programmes in the neo-patrimonial cases

work in villages that do not agree to the terms. This type of selective approach would likely be problematic for government agencies with constitutional responsibility for delivery of universal services. Overall, the support approach is frontloaded with extensive capacity building, in addition to significantly better service access standards, during implementation but on-going support is limited.

These two case studies show the close partnership or affiliation that can emerge between NGOs and government. They also show that NGOs can be good at providing more adaptive solutions to communities through working with alternative water sources as in Madhya Pradesh. Similarly, the greater autonomy of the NGOs mean they be more selective in who they can work with but this leads to questions over whether more intensive or selective approaches of NGOs can be followed by PHEDs at scale.

Service levels in the neo-patrimonial cases

This section now presents the household survey data on service levels achieved for each case. The data shows that all the programmes have delivered full coverage for improved water supply and either full or partial coverage for piped water supply. These coverage rates are more advanced than the state averages supporting the assertion that the case studies reflect (at least) relatively successful examples of community management. Yet despite the good coverage rates the service-level data indicates that much of the population continue to receive inadequate service levels. As shown in Table 6.2, the only two case studies where the programme villages have an average service level that meets government norms are the Odisha (Gram Vikas) and Chhattisgarh (PHED) cases. All other programme villages have an average service level of either 'no service' or 'sub-standard'. On a general level, the data suggests that service levels rarely meet government norms. This pattern of good access figures but poor service levels has been found in other studies of rural water supply programmes (Clasen, 2012; Godfrey et al., 2011) indicating that it is not unique to this context. In India the findings have mixed implications for the policy targets concerning piped water supply as they demonstrate that community management can deliver higher coverage rates of household piped water supply but, worryingly, even when this type of service is provided the level of service that

Table 6.2 Descriptive statistics on household service-level data

Case	State	Service level (median)	Percentage of population reaching basic or above service level
1	Jharkhand	Sub-standard	31%
2	Madhya Pradesh	No service	35%
3	Odisha	High	69%
4	Chhattisgarh	Basic	55%
6	Rajasthan	Sub-standard	19%

is achieved is not necessarily high. It is only in the Gram Vikas case in Odisha that over 65 per cent of those surveyed in programme villages reported high or improved service levels. This compares favourably to the other cases in this chapter, including the PHED-managed ones in Jharkhand, Chhattisgarh and Rajasthan and the NGO-Public Partnership case from Madhya Pradesh, where acceptable service levels range from just 19 per cent to 55 per cent out of those surveyed. Overall this section demonstrates that the neo-patrimonial case studies exhibit notable improvements through community involvement in terms of coverage rates compared to state averages but variable outcomes in terms of service levels. It was also possible to verify that four out of five cases had higher levels of success in the programme villages than the control villages.

Resources dedicated to support

This section now describes and analyses the resources dedicated to support and service provision in each case study. It first discusses the initial investment through capital expenditure (CapEx) and then goes on to discuss recurrent costs. Table 6.3 provides the key data here. It shows that it is only in two case studies where community contribution to CapEx were found with these being the NGO-supported Odisha case and the Swajaldhara example from Rajasthan. It is particularly interesting to note the differences between the government-supported Rajasthan case and the other PHED cases from Chhattisgarh and Jharkhand. As part of the Swajaldhara programme (2003–2009), the Government of India put much greater emphasis on ideas about demand-responsive community management compared to the more recent NRDWP (2009–2015). In the latter NRDWP the focus on cost-recovery is less emphasised and, as such, it can be said that community contribution towards CapEx is no longer considered a *necessary* step in rural water supply programmes (although it is still encouraged).

Across the neo-patrimonial villages there is variety in the recurrent costs of services. Odisha and Chhattisgarh, with the highest overall service levels, have the lowest recurrent costs per person. The contribution of support to recurrent costs

Table 6.3 Financial details for neo-patrimonial case studies

Case	State	Capital expenditure (CapEx, per person)	Percentage support contribution to CapEx	Annual recurrent costs (per person)	Percentage support contribution to recurrent costs
1	Jharkhand	$208	100%	$6	71%
2	Madhya Pradesh	$166	100%	$10	3%
3	Odisha	$169	82%	$4	10%
4	Chhattisgarh	$112	100%	$3	26%
6	Rajasthan	$93	89%	$11	15%

is also variable. However, at a general level, the evidence here shows that communities are able to contribute the majority of OpEx in these challenging cases but that 'top-up' subsidy is still required, which goes against the early principles of the Swajaldhara and broader DRA-community management model that prescribed communities contributing 100 per cent of OpEx costs (even if this was rarely achieved). When calculating the recurrent costs, capital maintenance costs were only available from a limited number of villages. They were available in three of the Rajasthan villages, which most likely reflects the longer operating time of those systems, as well as two of the Jharkhand villages. The costs ranged from around $0.5 to $1.5 per person in Rajasthan where committees have invested in new motorised pumps as back-ups or replacements. In Jharkhand the CapManEx costs are much lower between $0.15 and $0.20 per person but were also used to purchase new motorised pumps. Most tellingly though is the lack of CapManEx data that was found across the other case studies. This is thought to reflect a general lack of investment in CapManEx in these cases and a reliance on access to 'emergency' funding from government or other support entities when infrastructure fails. While there is nothing wrong with this type of support, it is (usually) more cost effective to pre-emptively manage repairs through periodic CapManEx investments. Either way, the research has not captured CapManEx in these cases either because it does not exist or because it is indirectly provided through government or other entity support and remains a 'hidden subsidy' within the broader costs of public administration.

Discussion of the neo-patrimonial cases

In this discussion section a number of key themes from this chapter are highlighted and further interrogated. These include considering what the shift in policy from Swajaldhara to the NRDWP has meant for government-supported community management, comparing the relative strengths and weaknesses of NGO and public support entities, and reflections on the extent and possibilities for the professionalisation of community management in neo-patrimonial contexts.

Beyond the demand-responsive approach

This chapter has helped illustrate the changes in Government of India policy toward community management through the Swajaldhara case in Rajasthan and the NRDWP cases in Chhattisgarh and Jharkhand. This change represents a shift away from the 'demand-responsive approach' to community management and move towards a model of direct provision with community involvement. The NRDWP states that 'decentralized, community-managed, demand-driven programme on broad Swajaldhara principles … will be encouraged. [However] Capital cost sharing by the community is left to the state to decide' (Government of India, 2013a, p. 9). In Jharkhand and Chhattisgarh the state governments have decided to reject a core assumption of the demand-responsive approach, that capital contribution either through community contributions of financial capital or labour is a prerequisite for a new scheme. This can be seen in the CapEx data with average community contributions of around 10 per cent of capital costs in Rajasthan but no contribution in the Chhattisgarh and Jharkhand villages. However, the shift can also be seen in the institutional structures for community management with a trend toward the 'formalisation' of community management institutions within the Indian state. The international origins of community management emerged at least partly due to development agencies trying to circumvent deficient (local) government systems (Harvey and Reed, 2006; Schouten and Moriarty, 2003). In India, however, as the Panchayat Raj Institution (PRI) decentralisation policies continue to be advanced community management appears to be becoming '(re)municipalised' into the local-self-government system of rural India. For example, in the cases given in this chapter, the Gram Panchayats in Chhattisgarh are now service providers while in Jharkhand VWSCs are established as a sub-committee of GP but have limited autonomy. They are also both reliant on direct and indirect government subsidies that cover significant proportions of operational expenditure. In comparison, the VWSCs in the Rajasthan case have greater independence and responsibility over decision-making and revenue generation. Yet, here, as they struggle with system augmentation they can perhaps more accurately be described as isolated. In conceptualising this change it is helpful to reflect on the 'ideal' accountability triangle for service providers that positions the state, service provider and users in relationship with one another.

Considering the growing body of literature that questioned the efficacy of the demand-responsive approach (Marks and Davis, 2012; Moriarty et al., 2013; Rout, 2014), it could be argued that that the shift away is an overdue change in policy. However, based on what has been presented in this chapter, the evidence on whether the NRWDP is more effective is still lacking and the policy clearly has its own problems in neo-patrimonial states. In both Jharkhand and Chhattisgarh, as per the NRDWP, the enabling support environment is meant to have two main 'institutional-threads' with one of these focused on hardware support through the PHEDs but the other focused more on software support through the Panchayat Raj institutions. In terms of direct software support to communities, the guidelines mandate the establishment of Block Resource Centres to provide support to communities, such as capacity building of the VWSCs (Government of India, 2013a). This research found no evidence that these institutions provided support to the villages or even existed in anything but name. Building on the ideas of Pritchett et al. (2013), these 'phantom institutions' suggest that the attempts to implement the NRDWP has led to (partial) isomorphic mimicry in terms of the enabling support environment for rural water supply. Within the overall political economy framework of the research, it is argued that it is more probable that such developments occur in the neo-patrimonial states than the other state groups and, in this sense, this case study reinforces this claim and raises queries over how to roll-out policy in these types of contexts.

Learning from scaled public health engineering departments and participatory NGOs

Another key difference across this chapter's cases has been the role of NGOs in two of the cases. To sharpen the comparison of the support entities and service providers a series of data-processing and analytical tools were developed which can help illustrate the differences and similarities here. These were designed to provide information on possible success factors for community management that had been identified from the wider literature. This includes the level of community participation in service provision and also an assessment of the level of partnering between the service provider and support levels for each case study. Table 6.4 below shows the level of community participation in the service delivery stage for each case study. It indicates that the two NGO-involved cases have higher levels of participation, ranked at the interactive level. This is because community members have more freedom to influence decisions through the VWSCs. For example, in the Gram Vikas approach all community members attended monthly meetings to take decisions with the VWSC on issues such as the duration of supply each day. For the government-supported programmes participation was ranked at the functional level, meaning that communities were given the opportunity to discuss service provider arrangements but had limited power to change them. The evidence corroborates similar observations that NGOs tend to work through a more participatory model than the government in India (James, 2004, 2011), although this can often involve more intensive community-orientated support activities that can limit scalability.

The NGO-involved programmes were also reported as having a broader partnering approach that is reflective of four out of six of the partnering models whereas the PHED-supported programmes were strongly related to the transactional model of partnering (see Figure 6.3). Transactional partnerships are

Table 6.4 Organisational data in the neo-patrimonial states

| Case | State | Organisational characteristics Professionalisation ranking (out of 100) | | Partnering typology | Participation in service delivery |
		ESE	CSP		
1	Jharkhand	55	67	Transactional	Functional participation
2	Madhya Pradesh	95	67	Transactional	Interactive participation
3	Odisha	75	61	Operational	Interactive participation
4	Chhattisgarh	50	39	Transactional	Functional participation
6	Rajasthan	50	33	Collaborative	Functional participation

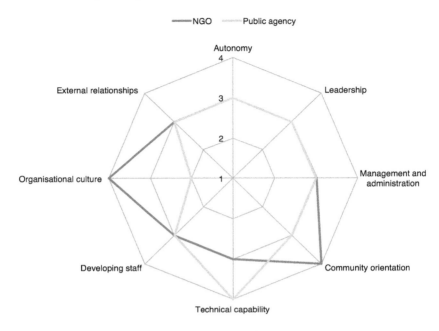

Figure 6.3 Subjective institutional assessment of NGO versus PHEDs in neo-patrimonial
states (1 = low capability; 4 = high capability)

characterised by structured negotiation followed by a clear division of labour
between partners for the remainder of the working relationship (Demirjian, 2002).
This type of partnership seems appropriate for government agencies working at
scale with many communities. However, these cases suggest it has trade-offs in
terms of lower levels of community participation compared to models governed by
a broader partnering approach. It is useful to reflect on the differences in the
approaches to support to help understanding why this may the case. For this
purpose, organisational theory can be used to contrast 'mechanistic' against
'organic' organisations (Burns and Stalker, 1961) as Franceys (2001) has done in the
context of urban water and sanitation providers. Mechanistic organisations follow
hierarchical structures of management and decision-making, and are best suited to
producing standardised goods or services at scale. PHEDs fit this description as
their core tasks involve infrastructure development and renewal as well as operation and maintenance of larger (bulk) water systems. Such tasks lend themselves
toward standardisation along mechanistic principles.

However, the task of providing support to communities, particularly with
regards to promoting community participation in water supply, is a more iterative
and creative process. Organic organisations are less hierarchical and therefore more
responsive and adaptive to context (Burns and Stalker, 1961), something that lends

itself to working closely with communities. It is thought that NGOs tend to be more organic and therefore have better participatory outcomes. Of course providing support for rural water supply will continue to require a large mechanical element, especially at the state-scale, so this should not lead to PHEDs moving to a completely organic model if they wish to promote higher levels of participation. There are mixed organisational types whereby mechanistic organisations have an organic component for the more creative elements of that organisation's work, such as research and development. This analogy has already been addressed in some ways by the prescriptions in the NRDWP that favour the establishment of Block Resource Centres that are separate from the core-structures of the PHED to lead on software support for these communities. However, as discussed previously, these have not become effective support institutions in the two cases from this chapter. This demonstrates an important lesson about the limits of institutional reform in such contexts: it is easier to design a 'good' policy than to implement it successfully.

With the notion that there needs to be a professionalisation of rural water supply to address inadequacies in the more voluntary models that had emerged in recent decades (Moriarty et al., 2013), an assessment of the professionalisation of service providers and support entities was made. At the support level it has also included an appraisal of the relative organisational strengths of the support organisations through an adapted WASH institutional assessment tool (Cullivan, 1988). Focusing on the professionalisation of the support entities the appraisal suggests that, on average, higher levels of professionalisation can be found in the two NGOs compared to the government agencies. Some of the key measures of professionalisation can be found across all cases. For example, the support entities had a formal mandate either as government agencies or approved by the relevant government agencies to work in the area. The professionalisation of the service provider was also assessed. This pointed to marginal differences between the cases but no clear pattern was found. Professionalisation was higher in Madhya Pradesh and Jharkhand because in these cases there was evidence of the systematic application of standardised guidelines for administration and complaint redressal, whereas the other cases had more *ad hoc* approaches based on personal relationships between community members, VWSC staff and junior engineer/officials from support entities. It is not possible to assess whether either approach appears to be associated with better outcomes in terms of service levels from just these five cases. The institutional assessment presented below helps to further characterise and differentiate the support entities. Again, it indicates a division between the NGOs and the PHED organisations. The biggest difference is that the PHEDs score comparatively higher on the technical capability whereas the NGOs have stronger measures on community orientation and organisational culture. This further enhances the arguments about organisational types made in the previous section.

Conclusion

This chapter has presented and analysed a series of case studies from the neo-patrimonial states in India. This included three cases where the PHED is the primary support entity and two cases that also have roles for NGOs either as partners or as the primary support entity. The findings from these case studies demonstrate that modest success can be found in these programmes, however it is the Gram Vikas case study from Odisha that stands out as the top performer from this chapter. The more participatory approach of this NGO case was contrasted with the standard working conditions of the PHEDs, which operate in a more bureaucratic manner. However, a discussion was had regarding the limits of transferring an NGO model into the public sector, while the shift in how PHEDs operate between the earlier Swajaldhara and later NRDWP programmes was highlighted. This trend indicates a trend away from the international principles of community management towards a model in which communities are not necessarily expected to contribute to capital costs and will receive on-going subsidy from the state. Together these case studies help illustrate the difficulties of delivering services in the challenging operating conditions of the neo-patrimonial states – modest success can be achieved but highly successful programmes require levels of input that are not considered scalable across the full government programmes, for the time being.

Note

1   With the idea that the daughters-in-law of the village are unlikely to move away once they move to the village as it is tradition for women to move to their husband's village when married and, more generally, men are more likely to migrate for work.

7 Community management in the 'social-democratic', middle-income states

This chapter presents and analyses the six cases from what could be called 'the sticky middle' of states where there is no clear narrative about them being the poorest, the richest or those shaped by the geographical contexts of hills and mountains. However, on closer inspection the cases here present a window into two particularly interesting facets related to rural water supply in modern day India. The middle-income states are where the most celebrated examples of social-democratic India can be found, that is, a development model that appears to be more focused on delivering human development outcomes rather than economic growth. The classic example being Kerala that has been characterised by a 'leftist' political economy (Ruparelia et al., 2011; Kohli, 2012) and has been able to deliver the highest HDI outcomes in the whole of India despite being only the ninth richest state. The middle-income states are also where the research has identified three of the most successful examples of donor-supported community management programmes in India, with the chapter presenting World Bank supported initiatives in Karnataka, Kerala and Punjab. It will be considered why internationally supported programmes appear to work especially well in these middle-income states as compared to the low-income and high-income states. There is also another case study from Kerala in the chapter that provides insight into the Kerala Water Authority (KWA) supported programme in Kerala. This case provides an example of community management through the public agency, the Kerala Water Authority, in the most decentralised state in India, as per the government's Devolution Index (Government of India, 2015b).

The final two case studies in this chapter are 'odd ones out' from the rest of the study. The case from West Bengal showcases the support provided by the NGO Water for People to communities managing handpumps in the saline intruded areas of coastal West Bengal. This is the only case study that focuses on handpumps. The other 'odd' case examines the community management of reverse osmosis plants and water kiosks for water delivery in a fluoride affected area of Telangana state. These final two cases provide additional insight into the role of community management with different types of technology, although as standalone cases it is recognised that the generalisability of the findings for these technologies will be more limited, as compared to the piped water supply from the other 18 cases. With three of the case studies from this chapter on World Bank supported programmes,

Table 7.1 Overview of the enabling support environment and community service providers

Case no. (inverse GDP rank)	Case name	State	Enabling support environment	Community service provider	Scale of support programme (approximate population)
7	Community-managed handpumps in Patharpratima, West Bengal	West Bengal	Water for People, Digambarpur Angikar (NGO), Gram Panchayat and the Jalabandhus	Water committee for each handpump	25,000
8	Decentralised drinking water service delivery – community-managed water purification units in Telangana★	Telangana	World Bank supported Jal Nirmal Project implemented by the Karnataka Rural Water Supply and Sanitation Agency	Village water and sanitation committee	3,061 villages
9	Jal Nirmal and beyond: supporting the community management of rural water supply in Belagavi district, Karnataka (World Bank)	Karnataka	World Bank supported Punjab Rural Water Supply and Sanitation Project implemented by the Department of Water Supply and Sanitation (Government of Punjab)	Gram Panchayat water and sanitation committee	7,260,000 (approximate coverage until 2014)
11	24/7 water supply in Punjab: international funding for local action (World Bank)	Punjab	World Bank supported Jalanidhi Project implemented by the Kerala Water Supply and Sanitation Agency	Scheme level executive committee	5,600,000 (approximate coverage of Jalanidhi supported programmes up until 2010; Baby Kurian 2010)
13	Jalanidhi programme in Nenmeni Panchayath, Wayanad district, Kerala (World Bank)	Kerala I	Gram Panchayat with support from the Kerala Water Authority	Beneficiary groups	45,723

Table 7.1 continued

Case no. (inverse GDP rank)	Case name	State	Enabling support environment	Community service provider	Scale of support programme (approximate population)
14	Community-managed rural water supply in Malappuram district, Kerala	Kerala II	Bala Vikasa	Water committee	5,000

Note: *In the Telangana and Andhra Pradesh case study report (reference) there is analysis of three different reverse osmosis schemes in Telangana and Andhra Pradesh. However, the Bala Vikasa case, which is from Karimnagar district in Telangana, is the only one with a community service provider (the other two are private-sector models), so only this part of the case has been included in the synthesis.

these will be described in an extended collective section. Then three shorter sections will focus on the Kerala II, West Bengal and Telangana case studies.

World Bank supported programmes

It is estimated that around $5 billion is allocated by the international investment banks annually to the global WASH sector. The World Bank is the largest of these funders and has historically provided significant support to India across a range of development sectors. It should be clear that in each of the cases that follow the enabling support environment agency remains the equivalent of a department within the PHED agency rather than being an actual World Bank body. What the World Bank support provides is additional capital which is tied to a mandate to work through a more demand-responsive, participatory community management approach than the supply-driven model of some state governments. The Bank also provides some training programmes for staff and, critically, conducts comprehensive monitoring of programme delivery to ensure the programme funds are being used for the prescribed purpose. Perhaps of most importance is that the World Bank support provides an impetus for greater flexibility and experimentation within the mechanical workings of the large, public bureaucracies that run rural water supply across the states. It should be noted though, and reflected on through this chapter and later in the investigation, that the most celebrated World Bank supported programmes have been in the middle-income states, with comparatively advanced decentralisation of public services and/or comparatively wealthy and well-educated rural populations. Going forward, the World Bank has provided investment of over $1 billion to work in four of the poorest states in India. Initial evidence from these programmes is that they have run into problems due to a lack of local government capacity.

Kerala Water Supply and Sanitation Agency (Jalandihi)

The Kerala experience with the World Bank supported Jalandihi programme started following the Cochin Deceleration of 1999 when policy-makers from across the states committed to the Sector Reform Pilot Projects that later (in 2009) became the Swajaldhara programme. In Kerala this lead to the state government committing to create the 'autonomous' Kerala Water Supply and Sanitation Agency to deliver a demand-responsive service delivery model in four districts of the state. The World Bank was an international advocate for the Cochin Deceleration and was therefore able to support state governments taking proactive steps to implement its principles. By 2001 it was supporting the Kerala Water Supply and Sanitation Agency in the Jalandihi Phase I programme which had a total funding of approximately $60 million to implement new schemes and establish community management systems for over 1.5 million people over a five-year period. Jalandihi Phase II has followed from 2011 to 2017 with an aim to scale the approach beyond the initial four districts to an additional 1.8 million people (Kerala State Planning Board, 2009). The World Bank committed over $220 million to this next stage with the explicit objectives to: '(i) support capacity building of sector institutions and support organisations; (ii) assist Government of Kerala (GOK) in implementing a state-wide sector development programme; and (iii) support project management costs' (ibid).

The case study focused on here comes from a scheme implemented as part of Jalandihi Phase I in Nenmeni Gram Panchayat, Wayanad district. With the size of Gram Panchayats in Kerala more akin to administrative blocks in other states, the Gram Panchayats have sufficient size to become a more fully professional local government unit. For example, in Nenmeni the GP has a population of nearly 50,000 people compared to a national average of circa 5,000. The Jalandihi support model can be described as front-loaded in that intensive support was provided during the capital expenditure phase that lasted 27 months starting in 2004. During this period a small multi-village scheme was rehabilitated and expanded to cover 18 of the 23 habitations and then a Scheme Transfer Memorandum was signed that transferred the assets to the GP. With the support of Jalandihi the GP then established a Management Transition Committee to begin the process of developing the community management capacity. This committee worked to develop eight beneficiary groups at habitation level to take on the operation and minor maintenance of habitation-level duties. These beneficiary groups also formed a Scheme Level Executive Committee to enable the coordination of activities and scheme-level operation and minor maintenance activities. Through this process a sophisticated 'professional community-based management' model was developed that operates until this day.

Karnataka Water Supply and Sanitation Agency (Jal Nirmal)

Karnataka takes the number two spot on the Government of India Devolution Index (Government of India, 2015c). It is not a classic example of a social-democratic state

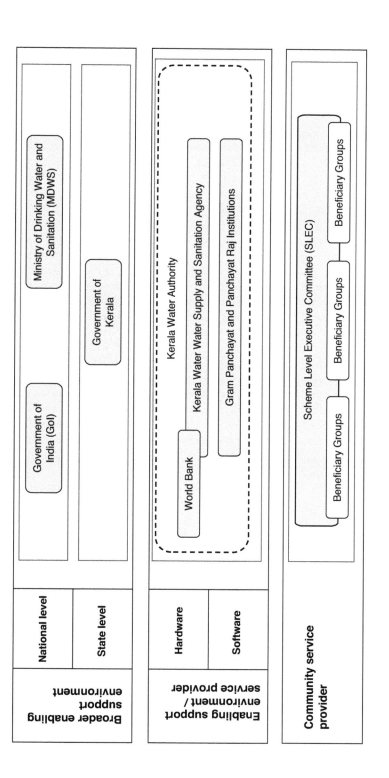

Figure 7.1 World Bank supported Jalanidhi programme of the Kerala Rural Water Supply and Sanitation Agency in Nenmeni, Kerala

in the same ways as Kerala but as Kohli (2012) argues all the South India states have greater social-democratic tendencies than the Hindi heartlands of the north. Together the extent of devolution and a relatively bottom-up approach to development means that in the researchers' views the state is more likely to be conducive to the demand-driven, decentralised service delivery promoted by the World Bank. Similar to the establishment of Jalandihi in Kerala, during the turn to community management in Government of India policy that occurred between the late 1990s and early 2000s, the Government of Karnataka was an active player who sought to implement the principles of the Cochin Deceleration within its rural water supply programmes. This led to the establishment of an autonomous project implementation body called the Karnataka Rural Water Supply and Sanitation Agency to take on the World Bank supported Jal Nirmal project.

The project aimed to provide 15.5 million people across 11 districts with new or rehabilitated water supply schemes and develop community management systems to take on the on-going operation and maintenance of the infrastructure. In total over two separate funding cycles than ran from 2002 to 2014 Jal Nirmal led to the establishment of 3,064 schemes across 744 Gram Panchayats involving $120 million of funding. The case study in Karnataka focused on the legacy of this programme by studying three villages that were part of Jal Nirmal and a control village that was supported as part of the general Rural Drinking Water and Sanitation Department. As Jal Nirmal has finished, the on-going support duties to the Jal Nirmal villages have been passed back to the department.

The three programme villages selected from Karnataka all have functioning VWSCs that have operated and maintained the water supply for over five years. Despite having considerably smaller Gram Panchayats than in Kerala, the villages here have also developed professional community-based management structures. In Shiraguppi village this includes having extensive office space, meeting rooms, facilities and employees at the GP level that can be used to support the VWSC management system. For example, these are used to record all VWSC meetings and telecast them so that people unable to attend the meetings in person can view the discussion. Such a commitment to transparency is also reflected in the publically available accounts that are published quarterly. It is such capacity at the GP level which is characteristic of states where the World Bank approach to community management has proven to be so successful. In multi-village schemes, such as Hirenandi, there are also institutional structures so that Joint Committees can be formed to take on the management of small-scale multi-village schemes.

Punjab Rural Water Supply and Sanitation Project

As opposed to Kerala and Karnataka, the state of Punjab scores weakly on the Devolution Index. Yet this has not inhibited its highly successful World Bank supported Punjab Rural Water Supply and Sanitation Project. Perhaps reflecting its greater tendency for centralisation, the Punjab Rural Water Supply and Sanitation Project has been implemented along a principle of a Sector Wide Approach (SWAp) meaning the principles of the project, such as mandating community

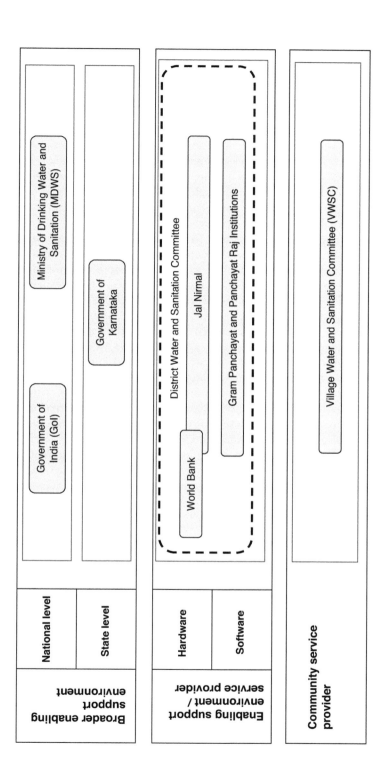

Figure 7.2 Jal Nirmal programme in Karnataka

contributions to CapEx, are being replicated across the whole state. With this project starting in 2006, the Punjab experience in many ways shows the next generation of World Bank supported projects as compared to Kerala and Karnataka.

The SWAp approach helps navigate a key barrier, which is the inconsistent approach of state agencies when supply-driven models are run alongside project units that follow a demand-driven model. In such cases, communities can question why one village is mandated to pay for services when other villages receive services free of charge (even if these services tend to be of lower quality and lack sustainability). Equally, the level of service supported by the Punjab Rural Water Supply and Sanitation Project is extremely high with an aim of supplying 70 lpcd and some villages now managing their own 24/7 piped water supply, a feat considered unrealistic in rural India up until recently.

The Department for Water Supply and Sanitation, Government of Punjab, established new project cells to take on the implementation of the project, while retaining its conventional structure for the continued support of pre-existing schemes (the SWAp applies to new schemes only). These project cells contain both technical teams and social mobilisation teams at all levels yet the key features of success for the village-level management comes from lowest forms of support. In particularly, the project mandates that junior engineers from the department sit on the community committees. This appears to work extremely well as it provides communities with direct access to support staff and technical knowledge within discussion. The naming of the committee as the Gram Panchayat Water Supply and Sanitation Committee (GPWSC) also makes explicit the sub-committee status of the committee within the Gram Panchayat system. This is thought to bring additional clarity as the constitutional mandate for Gram Panchayat management and then establishing Village Water and Sanitation Committees with various levels of autonomy from the GP, provides a point of confusion in the implementation of the NRDWP. In all the villages from this case study, the GPWSCs managed single-village schemes that provide household water supply to the population.

Decentralised Kerala supported community management

The 'other' Kerala case study comes from Malappuram district and showcases the advanced decentralisation that has happened across the state. Following the constitutional reforms of 1993 and 1994 to advance the decentralisation of public services, Kerala began the People's Planning Campaign in 1996 that devolved 33 per cent of state funds to the PRI to spend on development plans that were developed locally by Gram, Block and District Panchayats (Heller et al., 2007). Although the People's Planning Campaign faced some problems, it represents one of the earliest and most comprehensive examples of state devolution in India (ibid.) which is reflected in Kerala's continued presence as the most decentralised state, as per the Government of India devolution measurement index (Government of India, 2015c). In Malappuram the GP has a population of approximately 45,000 people and fulfils the on-going service support functions to a beneficiary group at the habitation level that takes on the service provision. The centralised Kerala

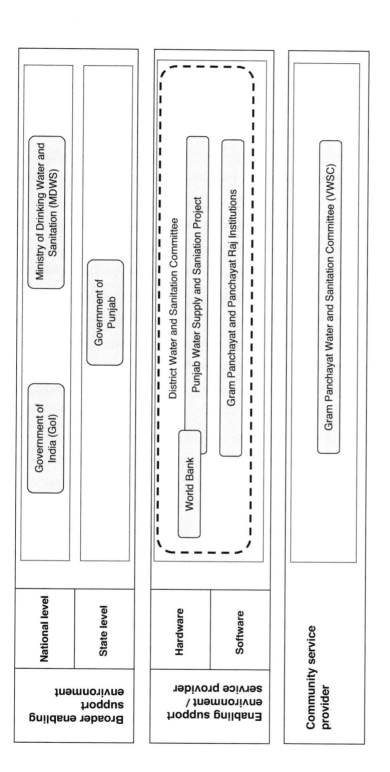

Figure 7.3 Punjab Rural Water Supply and Sanitation Project

Water Authority provides the role of implementation agency developing new infrastructure but then passing on the infrastructure to the beneficiary group and support functions to the Gram Panchayat. This represents one of the only examples where the local government agency becomes the principle support agent for community management, which is an arrangement common in many other parts of the world, such as in African countries such as Ethiopia and Latin America countries such as Honduras (Lockwood and Smits, 2011).

The managerial process for rural water supply within the local government system reflects a highly professional decentralised institutional set-up. New schemes proposed by the Beneficiary Group are passed to a ward-level Gram Sabha (public meeting with representatives from approximately 500 households). Upon approval in this meeting the request is passed to the GP who considers the request and passes a tender to the Kerala Water Authority for technical planning. The Gram Panchayat then approves the plans and has to mobilise the funds via its devolved funding stream. The Kerala Water Authority executes the work but the GP oversees and quality assures the work through its own fulltime public works engineer. The Beneficiary Groups then take on the on-going operational and minor maintenance with on-going support from the Gram Panchayat. The Beneficiary Groups have a similar set-up to the conventional VWSC from other states, with positions for President, Secretary, Vice-President and Joint Secretary. However, rather than being constituted as sub-committees of the Gram Panchayat, in Kerala they are registered under the Societies Act (1890) as independent bodies for charitable purpose. In this sense, the case reflects a form of self-mobilisation, as per the participation ladder, in that an autonomous community group 'takes responsibility for administration, management and operation and maintenance, either directly or by outsourcing these functions to external entities'.

Handpumps in West Bengal

The Government of India is moving beyond handpumps to piped water supply but the case study from West Bengal provides an example where they still play a critical role in villages. In Patharpratma block, which sits on the coast line of the state, there is a serious issue of saline intrusion in local aquifers making many boreholes obsolete. Water for People, the international NGO, has been working with government agencies to try to overcome this problem through what it calls a 'programmatic approach' to addressing water and sanitation needs across the administrative block. It supports the implementation, operation and maintenance of handpumps with deep bores that provide potable drinking water to the population. As will be reported later, the actual service levels delivered from this infrastructure are limited and the community continue to use traditional ponds for much domestic water needs, apart from drinking and cooking which comes from the handpumps. In this sense, the case reflects only a marginal success by Indian standards but it is of value to this study for two reasons. One because it provides insights into the role of community management in responding to water insecurity when conventional public provision has failed and, two, due to the continued

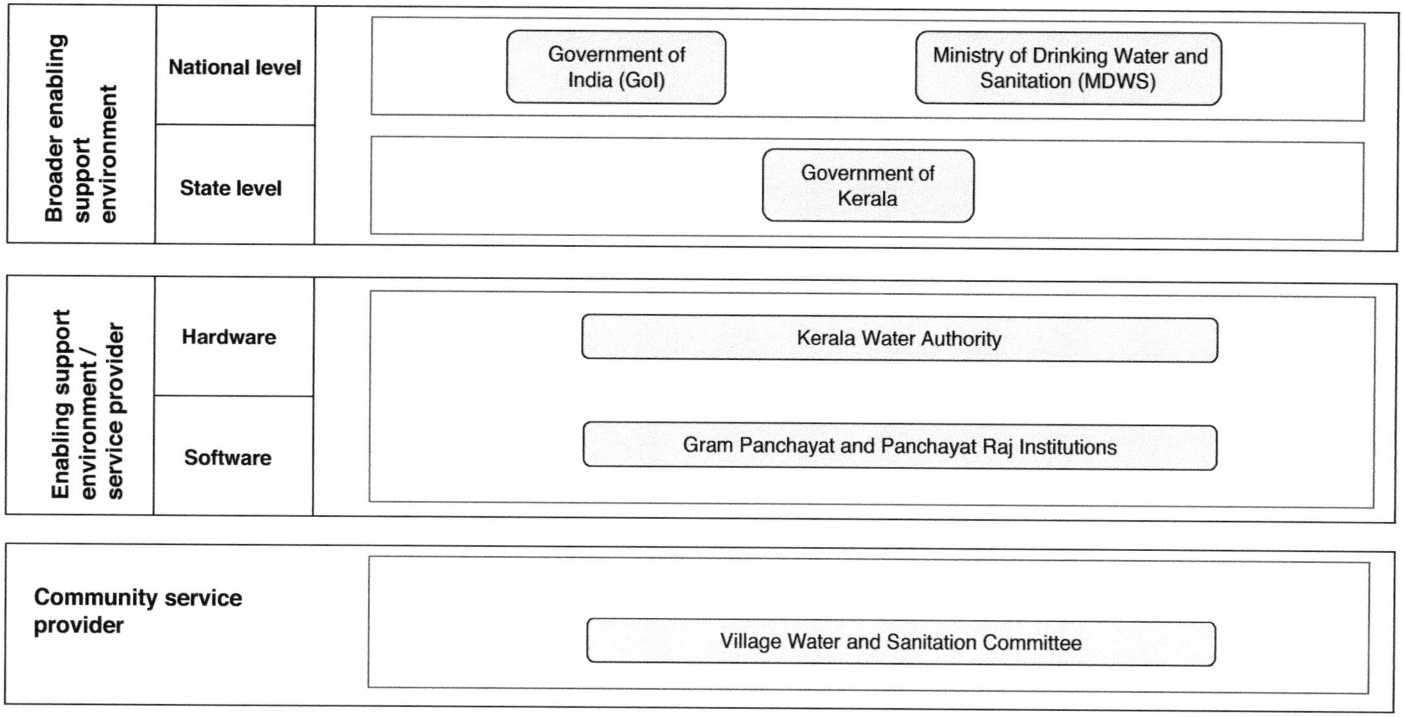

Figure 7.4 Kerala Water Authority case

Broader enabling support environment

- National level — Government of India (GoI) — Ministry of Drinking Water and Sanitation (MDWS)
- State level — Government of Kerala

Enabling support environment / service provider

- Hardware — Kerala Water Authority
- Software — Gram Panchayat and Panchayat Raj Institutions

Community service provider

- Village Water and Sanitation Committee

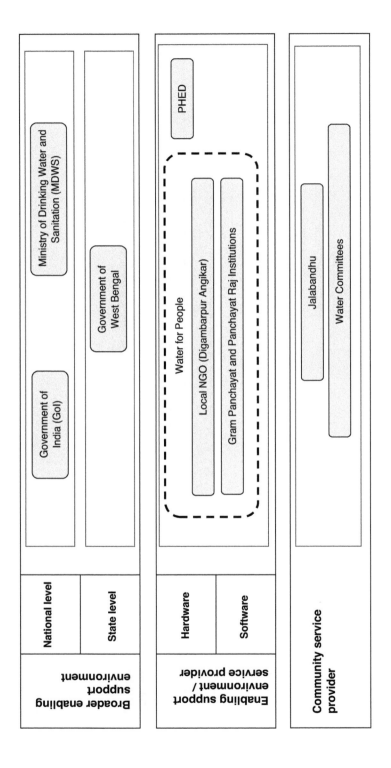

Broader enabling support environment

| National level | | Government of India (GoI) | Ministry of Drinking Water and Sanitation (MDWS) |
| State level | | Government of West Bengal | |

Enabling support environment / service provider

| Hardware | | Water for People | Local NGO (Digambarpur Angikar) | PHED |
| Software | | | Gram Panchayat and Panchayat Raj Institutions | |

Community service provider

| | Jalabandhu | Water Committees |

Figure 7.5 Supporting handpumps in West Bengal

importance of handpumps in many other parts of the world, particularly Africa, it may enable the study to offer insights into these wider contexts.

Water for People has operated in the area since 2006. In this case study, the focus is on Digambarpur Gram Panchayat where Water for People supports local NGO, Digambarpur Angikar, as well as the Gram Panchayat itself. The service provision tasks are then shared across three organisations, with the Gram Panchayat providing funding and major maintenance and an unregistered water committee doing the basic operation and minor maintenance, often working on a single water point. The technical maintenance is the only professionalised element whereby the water committees outsource repairs services to private handpump mechanics called Jalabandhus, who cover a number of schemes. The role of Water for People is to provide capacity building and advice to the enabling support environment of local entities. It does this by training the staff of the local NGO, it also helped train and establish the network of Jalabandhus and it advocates good water management principles at the Gram Panchayat level. This set up is not found in the neighbouring Gram Panchayat, Dakshin Gangadharpur, where the basic government system operates and which was taken as the control village.

Additional potable water treatment model in Telangana

The challenges of chemical contamination in drinking water schemes, built by the then Andhra Pradesh rural water department, led to various interventions by NGOs and the private sector. In parallel to the on-going, often household piped, water supply managed in the ways described for other case studies in this chapter, with varying degrees of community involvement, small village treatment plants were constructed through external interventions so that households could obtain approximately 3–5 litres of potable drinking water per person per day in addition to the piped supply continuing to be used for laundry and bathing. One such programme was developed by Bala Vikasa which began as an NGO in 1977 following a holistic community development model in Andhra Pradesh (which before the bifurcation of the state in 2014 included Telangana as well as the current state of Andhra Pradesh). As the magnitude of fluoride-related health issues emerged across the state, including high rates of skeleton and dental fluorosis in rural areas, the NGO committed to working on the expansion of community-managed water purification plants in 2002. This led to a work programme supporting the construction of reverse osmosis plants and the establishment of water kiosks to sell high quality potable water for drinking and cooking.

Bala Vikasa position themselves as working through a community management model as opposed to the many commercial reverse osmosis systems that have been developed in Andhra Pradesh and Telangana over the past two decades and which also follow a water kiosk model. The community was therefore being asked to run its own piped scheme, through the VWSC as well as being involved in some level of communal management of the additional potable water scheme. Bala Vikasa provides overall coordination and oversight of the institutional system. It has to receive an application from a village that they are interested in developing a reverse osmosis plant; it then holds meetings in the village to explain the process. It will begin work constructing the plant only when 80 per cent of households agree to take on the equal ownership and the village is able to raise 60 per cent of the capital costs, which can be taken as cash but also in the form of labour and land for the site. Bala Vikasa then commissions the construction of the reverse osmosis plant from private-sector contractors and passes it over a newly established water committee for the on-going operation and maintenance. The water kiosk distribution system operates through an 'ATM model' whereby community members can access water 24 hours a day using pre-paid cards from an automated machine attached to the reverse osmosis plant.

Recognising the additional technical burden of such complex technology the water committee is linked with a private-sector technical support unit called Innovative Aqua Systems Private Limited that provides regular technical monitoring and minor repairs for a set fee from the committee. The water committees are effectively informal community bodies, not registered under the Societies Act or under the Gram Panchayat, however Bala Vikasa seeks to promote some degree of formality. It provides written mandates of the role of each member of the water committee, written in local languages, and then provides specific training to committee members following their nomination onto the committee by the wider community.

Service levels

Assessing the service levels across the cases it is clear that those cases achieving improved or high service levels in the programme villages are where there is the most prevalent use of household connections. In both the Kerala cases and Punjab 100 per cent of those surveyed reported having a household connection and over 90 per cent achieve a score of either improved or high services. In the Karnataka case, households were more likely to report having lower service levels due to the use of public stand-posts. The cases from West Bengal and Telangana provide useful comparators to each other. In both cases potable water is provided through what could be described as supplementary systems. In Telangana the service levels reflect

Table 7.2 Service levels in the social-democratic case studies

Case no.	State	Service level (median)	Percentage of population reaching basic or above service level
7	West Bengal	No service	0%
8	Telangana	Improved	87%
9	Karnataka – World Bank	Sub-standard	37%
11	Punjab – World Bank	High	98%
13	Kerala I – World Bank	High	100%
14	Kerala II – Local self-government	High	94%

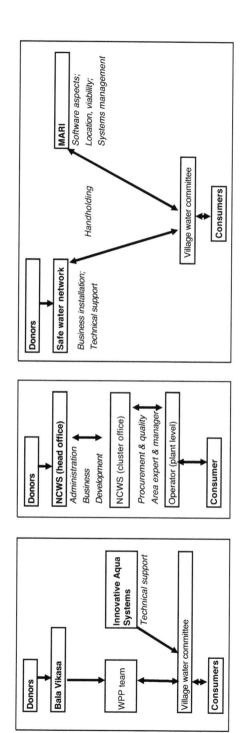

Figure 7.6 Institutional models for water kiosks in Andhra Pradesh and Telengana

the combined level of service of the potable water kiosk supply for drinking and cooking and the non-potable supply from the publically managed piped network for other domestic uses. Together these provide a level of service that has been reported as improved or high on the service level ladder for the majority of households as per the household surveying. However, in West Bengal, the handpumps provide the crucial potable source of water for drinking and cooking but households then rely on local traditional ponds for other domestic uses of water. Use of this unimproved secondary source was not captured via the household survey tool. Overall this means that the households in West Bengal receive only very small amounts of water from improved water sources (on average 12 lpcd) which is far below the norms of 40 lpcd.

Resources dedicated to support

The costs of services across these cases illustrate the marked difference between the West Bengal case study and the rest. In West Bengal the case is a handpump only case study with the capital and recurrent costs reflecting the lower costs of this type of service. Across the other case studies, the CapEx costs range from $184 to $282 for various forms of piped supply with the highest cost found in Telangana for the reverse osmosis plant that comes to $279 per person. In all the middle-income case studies, the communities are making at least some level of contribution to CapEx although this ranges from 2 to 17 per cent. Recurrent cost range between $12 and $50 when excluding the handpump case study. This is a considerable range between programmes demonstrating that there is little standardisation across the different states. Similarly, the range of contribution for the recurrent costs from communities ranges from covering 96 per cent of these costs in Karnataka to less than half the costs in Kerala I – World Bank, Telangana and the West Bengal case study.

Table 7.3 Financial data for social-democratic case studies

Case no.	State	Capital expenditure (CapEx)	Percentage support contribution to CapEx	Annual recurrent costs	Percentage support contribution to recurrent costs
7	West Bengal	$38	98%	$2	77%
8	Telangana	$279	88%	$16	59%
9	Karnataka – World Bank	$282	95%	$12	4%
11	Punjab – World Bank	$247	94%	$50	46%
13	Kerala I – World Bank	$184	83%	$30	52%
14	Kerala II – Local self-government	$221	97%	$32	30%

Discussion and further analysis

Institutional analysis is central to this research in answering the research questions regarding 'what' type of supporting organisations are found in successful community management programmes, but also there are important questions regarding 'how' better support organisations can be developed in the future, both in India and elsewhere. This chapter contains the case studies that showcase three major processes of World Bank supported reform in rural water supply. Donor-led public sector reform processes have been a key process in development in the twentieth century. Since the World Bank and IMF advocated structural adjustment programmes in the 1990s that led many African countries to introduce far-reaching market liberalisation policies that were damaging (at least in the short term), there has been scepticism in the development studies literature regarding donor-led reform. However, despite this scepticism, multilateral and bilateral donors are extremely important actors in the development institutional ecosystem, providing much needed capital for investment but also knowledge and skills to drive positive change. In the Indian rural water sector it is noted that in the James (2011) review of four successful community management programmes across India, three of these were the World Bank supported programmes in Kerala, Karnataka and Maharashtra. In this sense, the wider literature indicates that donor-support programmes are where some of the most successful examples are found.[1] The evidence from this chapter reconfirms this finding with the World Bank supported programmes providing the cases where the highest service levels have been found. The institutional analysis tools have also found high levels of performance within the World Bank supported rural water supply agencies. As shown in Table 7.2 and Figure 7.7, these agencies are rated as having strong performance across the board, including both technical capability and community orientation. This compares favourably to the other types of support in this chapter, including the Gram Panchayats and the NGOs, and even more favourably compared to the conventional PHED institutions, assessed in other chapters. This suggests that the World Bank support provides the initiative for successful reform in the public agency support of community management.

The role of the World Bank in driving institutional reform is particularly interesting as on a simplistic level it goes against some of the literature that suggests meaningful reform tends to be domestically driven and locally owned (Andrews et al., 2013). Yet this is misunderstanding the role of the World Bank in such reform programmes, as the Bank is not the primary driver of any reform. In Kerala and Karnataka it was the state officials that sought World Bank support following the Cochin Deceleration in 1999 as they understood it as an opportunity to access more funds. The World Bank's role was to then place conditions on the funding to shift towards more demand-responsive and participatory modes of service delivery then, crucially, they monitored the programmes against the mutually agreed upon principles. This then meant the public agencies could create independent project departments that can work under these rules which were akin to 'experimental spaces' beyond the standard guidelines and operating procedures of the mechanical

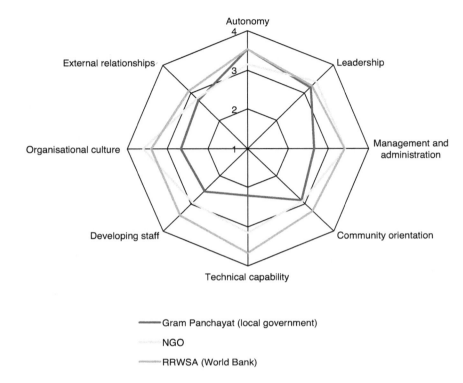

Autonomy

Leadership

Management and administration

Community orientation

Technical capability

Developing staff

Organisational culture

External relationships

——Gram Panchayat (local government)

NGO

——RRWSA (World Bank)

Figure 7.7 Enabling support entity institutional analysis in the middle-income cases (mean score by enabling support entity type)

public agencies. In short, they effectively allowed what were described as organic organisational types in the previous chapter, to emerge within the larger public bureaucracies.

What is interesting in comparing these two early cases to the later Punjab case study is that lessons have clearly been learnt about the limitations of this approach in terms of creating two-tiered rural water supply agencies. In Kerala and Karnataka project units were developed within the rural water supply public agencies to deliver the World Bank programme and these had to work on different conditions to the standard operating approach of the rest of the public agency. This then led to problems when villages were able to play off the different parts of the public agencies against one and another. For example, a village would question why it had to pay tariffs in the World Bank programme but not in the standard government one. In the later Punjab programme this was rectified by the state government adopting a Sector Wide Approach (SWAp) to the financing of rural water services. SWAp has been applied widely in developmental projects since at least the 1990s particularly in the health care sectors (Peters and Chao, 1998; Peters et al., 2013), but the application of such an approach to donor-supported programmes in the water sector, at least in India, appears limited. The SWAp approach has the added benefit that the changes become more widely institutionalised across the whole agency and staffs do not consider the practices as related to the project so therefore only temporary measures. In Punjab, which comes out in the bottom quartile in terms of the Devolution Index, as compared to Kerala and Karnataka that come out at number 1 and 2 (Government of India, 2015b), respectively, it may also be that having a centrally imposed policy framework is conducive with the working approach of the public administrative system.

Before moving on it should be considered that the conditions placed by the World Bank on the public agencies reflect the ideals about rural water supply that are common within the international literature, such as favouring demand-responsive community management. In some ways providing conditional aid in this manner could be considered imperialist and ideological, especially in light of limited evidence on whether such models are actually more effective than supply-driven approaches. Yet in these middle-income states the state governments are strong enough to not merely adopt the principles outlined by the Bank but work together to adapt them to context. In the Punjab this can be seen with the shift from the principle that communities should contribute 10 per cent of capital costs to a standard fee of INR400 per household (or INR200 for below-poverty line households) to capital costs. This was deemed an appropriate level in terms of affordability and as it was standardised it has become commonly accepted across the state. This is the type of iterative adaptation that is considered critical to having

Table 7.4 Organisational data from the middle-income case studies

Case no.	State	Organisational characteristics (summary indicators)			
		Professionalisation (out of 100)		Partnering typology	Participation in service delivery
		ESE	CSP		
7	West Bengal	30	33	Operational	Interactive participation
8	Telangana	90	75	Operational	Interactive participation
9	Karnataka	100	89	Transactional	Interactive participation
11	Punjab	75	78	Collaborative	Self-mobilisation
13	Kerala I – World Bank	70	100	Transactional	Self-mobilisation
14	Kerala II – local self-government	100	64	Operational	Interactive participation

successful reform processes within specific states. However, linking this back to the arguments made in Part I about the socio-economic carrying capacity of states and also the political economy of states, it is felt that these middle-income states have the capacity to undertake such processes and this may be lacking in poorer states. Moving forward the World Bank has recently provided large loans to some of the poorer states in India and the early anecdotal evidence on the success of these programmes is mixed. It is contended that while donor support can provide a catalyst for reform it is only likely to work in states that have the required institutional capability to absorb the additional capital and create new public institutions where it will continue to work.

Conclusion

The case studies from the middle-income states were labelled the 'sticky middle' in the introduction to this chapter reflecting the variety of cases selected from these states. However, they did contain the three World Bank supported programmes from the research which helped illustrate how donor support can lead to the reform of public agencies to become effective support entities for community-managed rural water supply. Similarly, through the West Bengal and Telangana case studies they showed the role that NGOs can play in supporting community management in difficult operating conditions. These examples are considered to reflect the flexibility of the community management model in terms of different forms of support.

Note

1   Although it is noted that there may be a systematic bias towards donor-supported programmes as due to funding requirements they are more likely to have systematic documentation of the programme, making any success more visible to (international) researchers.

8 Community management in the 'developmental', high-income states

Some of the major economic heartlands of India are in Maharashtra, Gujarat and Tamil Nadu. These states have annual GDPs of over $100 billion and represent three of the top four states in terms of absolute economic wealth, with only Uttar Pradesh buoyed by its disproportionality large population of over 200 million people also having a GDP of over $100 billion (Planning Commission, 2014). They also represent the three richest states per capita in this study, barring the economically prosperous but small Himalayan state of Sikkim (Reserve Bank of India, 2015). Although the economic history of each state is distinct, they share similar geographies that include substantial navigable coastlines making them historic centres of trade. Fitting the model of developmental states proposed by Kohli (2012), the states still have large populations that remain in relative poverty, especially in rural areas. This mix of economic wealth and modest human development has been linked to a governance approach where a strong state government seeks to maximise the competitive advantage of the state and drive economic growth in a top-down fashion but which leaves some sections of the population behind.

It can be argued that this mode of top-down development is reflective of, or at least analogous to, a major trend in the rural water supply sector within these states, which is a technical shift away from single-village schemes to the more sophisticated multi-village schemes. These multi-village schemes provide a possible trajectory of development for many regions in India as numerous policy documents have emphasised the importance of shifting away from groundwater to surface water schemes (Government of India, 2012b, 2013a). However, due to the higher technical demands of managing surface water, including the necessity for water treatment, it often becomes more appropriate to design and manage schemes at scales much larger than single villages, which is driving the trend towards multi-village schemes. The multi-village schemes discussed in this chapter cover hundreds or even thousands of villages and this scale presents a fundamental challenge to the principles of community management – that communities can operate and maintain their own water infrastructure. However, there remains a distinction between the management of bulk water infrastructure and local-distribution systems which means there are still ways to link community management into multi-village schemes. This chapter reveals this can take different forms that can either retain many of the core principles of community management or move towards what can be described as a 'utilitisation' of rural water supply, whereby communities become simply consumers in the same way as much of the urban population is. Although five cases are presented, there are two sets of 'twin-cases' from the same state. These are the cases from Gandhinagar and Kutch in Gujarat which showcase the support provided by the WASMO and the cases from Morappur and Erode in Tamil Nadu that show two related but distinct ways the Tamil Nadu Water and Drainage Board (TWAD Board) provide support to villages. These pairs of cases are presented together. The other case is from Maharashtra and

Table 8.1 Overview of the enabling support environment and community service providers

Case no. (inverse GDP rank)	Case name	State	Enabling support environment	Community service provider/ community organisations	Scale of support programme (approximate population)
15	WASMO in Gandhinagar District	Gujarat	WASMO	Pani Samiti (water committee)	16,000,000 (total WASMO coverage)
16	WASMO in the desert Kutch region	Gujarat	WASMO	Pani Samiti (water committee)	16,000,000 (total WASMO coverage)
17	TWAD Board and the Panchayat Raj Institutions in Erode district	Tamil Nadu	TWAD Board and the Rural Development and Panchayat Raj Department	Gram Panchayat	39,000,000 (total TWAD Board coverage)
18	TWAD Board and the Hogenakkal Water Supply and Fluorosis Mitigation Project in Morappur district	Tamil Nadu	Hogenakkal Water Supply and Fluorosis Mitigation Project and the Panchayat Raj Department	VWSC as a sub-committee of the Gram Panchayat	3,300,000 (Hogenakkal Water Supply and Fluorosis Mitigation Project)
19	Maharashtra Jeevan Pradhikaran (MJP) and the Shahnoor Dam project	Maharashtra	Maharashtra Jeevan Pradhikaran	Water committee as a sub-standing committee of the Gram Panchayat	50,000 (Shahnoor Dam multi-village schemes)

focuses on the support provided by the Maharashtra Jeevan Pradhikaran (MJP), a public utility, in Amravati district, through a multi-village scheme tied to the Shahnoor Dam project.

Water and Sanitation Management Organisation and developmental participation in Gujarat

The history of WASMO can be traced back to the Ghogha Regional Rural Water Supply Project in Bhavnagar district, which was supported by the Royal Netherlands Embassy from 1997 to 2001 (Das, 2014). This project provided a working example of a successful community management model in Gujarat and encouraged by the policy shift to community management through Swajaldhara at the national level, the Government of Gujarat instigated an institutional reform process in its water sector to move towards this model (James, 2011). This led to the formation of WASMO as a Special Purpose Vehicle in 2002 to lead on rural water supply provision. As shown in the institutional map, it continued to work alongside the Gujarat Water Supply and Sewerage Board, the statutory body mandated to provide water and sanitation services across the state, as well as another newly formed body the Gujarat Water Infrastructure Limited. Both these latterly mentioned organisations remain hardware agencies with the Gujarat Water Supply and Sewerage Board an asset creation agency and the Gujarat Water Infrastructure Limited operating bulk water schemes across the state. In this sense, WASMO is a rare case from this study that has seen a highly participatory model of community management emerge from a largely domestically driven reform process, rather than due to World Bank support or NGO influence. As such a section at the end of this chapter considers the reform process in more detail so to identify whether there are lessons for other states.

The core mantra of WASMO is 'users are the managers of water supply' and, in this sense, the approach of WASMO is congruent with the decentralising PRI reforms. However, while the PRI reforms have been criticised for being inadequate because of the devolution of functions to lower levels of government have not been matched with the adequate passing down of funds and functionaries (Asthana, 2008; Johnson et al., 2005), this research suggests that WASMO has taken seriously the process of subsidiarity to Pani Samitis (water committees). The WASMO programme mandates these entities as the service provider but matches this devolution of responsibility with a significant capacity-building programme of village-level functionaries as well as empowering the committees to collect sufficient revenue from users to cover costs, in part through demand-generation activities.

Since 2002 support has been provided to over 18,000 villages through a three-cycle project approach. These cycles are now described based on the reporting in the case study reports (Chary Vedala et al., 2015a, 2015b). Cycle one is focused on community mobilisation and lasts usually three to six months. In this stage, social mobilisers with specialist training in participation appraisal techniques work in the villages. The staffing divide between these software staff and the

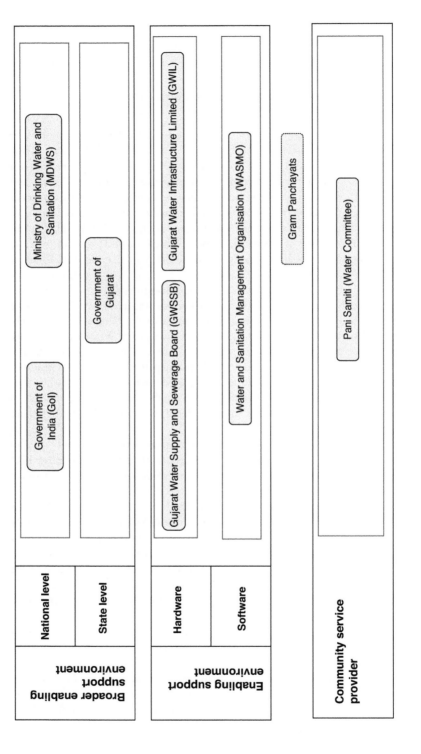

Figure 8.1 Simplified schematic of institutional arrangements for enabling support environment under WASMO in Gujarat

hardware staff is reported as 50/50. Such a high rate of social mobilisers has not been seen in any other public agency investigated in this study. Before entering cycle two, communities must agree to contribute 10 per cent of the capital costs of any assets created as part of the project and establish a Pani Samiti to help plan the scheme and take on its operation and maintenance going forward. As with VWSCs in other states, these bodies are an official sub-committee of the Gram Panchayat and must be elected by the Gram Sabha. The Pani Samiti is constitutionally empowered by a Government Resolution issued by the Gujarat government in 2002 and must have a separate bank account from the Gram Panchayat. By the end of cycle one, a Village Action Plan will be approved by the Gram Panchayat and in cycle two the physical implementation of the project is followed through. WASMO's role here is to provide technical assistance and monitoring but the Pani Samitis lead the projects, using private contractors. They must also set the water tariff charge in consultation with WASMO and the broader public. Following the asset creation, WASMO provides dedicated support for a further 12 months in the third cycle. This 'hand-holding' phase for service delivery includes further capacity building of the Pani Samiti staff as well as detailed monitoring of performance.

The decision to include two cases from the same programme was to assess whether the success of WASMO was the same in a core region of the state (i.e. Gandhinagar district) as a peripheral region with a more challenging environment (i.e. the desert region of Kutch where WASMO had be delivering rural water supply support in the context of reconstruction following the 2001 earthquake). However, the case study found the institutional support mechanisms and service provider arrangements were identical in both cases with Pani Samitis acting as the service providers. The technological set-ups for water supply were also similar with each village having a single-village scheme that is supplied by local ground water sources. However, it is notable that in Gujarat, although most villages effectively have a single-village scheme, in Gandhinagar and other core parts of the state it is common for the village systems to be also connected to the large-scale bulk water schemes for irrigation that run throughout much of the state. In practice, in the villages studied in this research, there was either no or minimum water drawn from these bulk systems apart from one village. However it is noted that this set-up clearly adds an additional level of resilience to the supply system in this water-stressed state.

Tamil Nadu Water and Drainage Board and the 'third way' public model

The enabling support environment for rural water in much of Tamil Nadu can be conceived of as a public–public partnership between the TWAD Board (PHED-type organisation) and the Rural Development and Panchayat Raj Department that supports the Panchayat Raj Institutions across the state. The set-up follows the NRDWP closely in that the TWAD Board is charged with asset creation for water infrastructure, which, in the case of single village schemes, is then passed over to the Gram Panchayat for operation and maintenance. In the service delivery phase the Block Development Office of the Rural Development and Panchayat Raj Department then largely supports the Gram Panchayat through provision of administrative training and as the nodal agency for the Gram Panchayat to access additional funding and technical support. For example, if the Gram Panchayat planned some additional capital works it would first go to the Block Development Office to get the plans sanctioned, which would then be passed to TWAD Board for technical implementation or oversight. The Gram Panchayats across both of the programme villages from the Tamil Nadu cases in this study have established VWSCs but the autonomy of the committees is limited and in essence the Gram Panchayats retain the role of service provider.

Both the cases were selected though as they had been part of a specialist programme from 2005 to 2007 named the Tamil Nadu Rural Water Supply and Sanitation Programme (TNRWSSP). This was a pilot programme that specified its aims as making TWAD Board engineers 'facilitators' of good water services rather than 'service providers' (Nayar and James, 2010). In practice this meant greater interaction between engineers and communities as well as training and a series of information, education and communication (IEC) activities before additional capital works were undertaken in them (i.e. new overhead tanks, extended distribution networks and boreholes). The villages were selected for the pilots due to water insecurity issues as it was deemed additional community participation would have the greatest chance of working and also maximum benefit in such contexts. This led to the establishment of VWSCs as sub-committees of the Gram Panchayat to take on the management of local water systems and help promote water conservation measures within villages. Although the pilot ended, the legacy of the programme still exists within the villages to some extent. In Erode this is particularly evident as it now receives just the standard PHED-PRI support for water supply. However, as the current President of the Gram Panchayat and VWSC is on his third term in office, he has built on his initial training with over ten years of experience. The VWSC meets less regularly than before, but as four of its members are also on the Gram Panchayat board they can advance the management of water supply largely through the Gram Panchayat itself. This is reportedly a typical set-up for Tamil Nadu – although a VWSC exists on paper much of the tasks are completed through the Gram Panchayat.

In Morappur the case study shows a more sophisticated enabling support environment that goes beyond the standard PHED model. In this area the water insecurity issues are more fundamental involving fluoride contamination of groundwater and a collapse in groundwater levels reducing the yield on many local sources. This situation has resulted in an additional level of support and investment in this block, including its selection as the area for one of 15 Government of India supported National Rural Drinking Water Security Pilot Projects (NRDWSPP) that ran from 2012 to 2014 (Ministry of Drinking Water Supply, 2013). This can be considered as a form of CapManEx on software support, as per the research framework, as this pilot meant further training and IEC activities focused on promoting water conservation, including for both domestic and productive uses

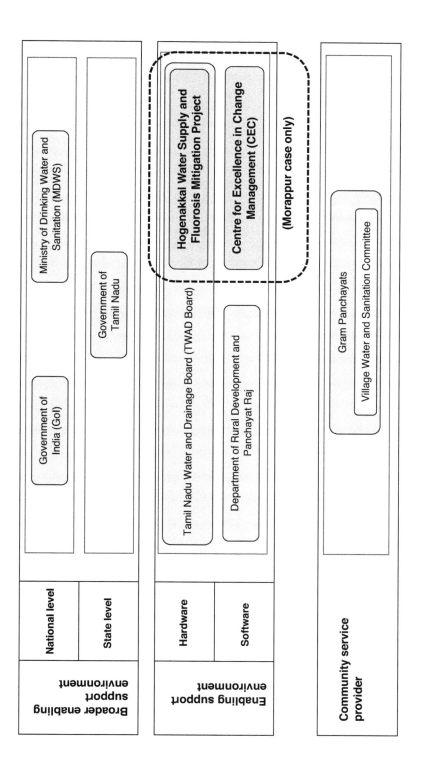

Figure 8.2 Institutional set-up of Tamil Nadu cases

Broader enabling support environment

- National level
  - Government of India (GoI)
  - Ministry of Drinking Water and Sanitation (MDWS)
- State level
  - Government of Tamil Nadu

Enabling support environment

- Hardware
  - Tamil Nadu Water and Drainage Board (TWAD Board)
  - Hogenakkal Water Supply and Fluorosis Mitigation Project
- Software
  - Department of Rural Development and Panchayat Raj
  - Centre for Excellence in Change Management (CEC)

(Morappur case only)

Community service provider

- Gram Panchayats
  - Village Water and Sanitation Committee

(i.e. promoting drip irrigation), were conducted. In practice, however, the impact of this software intervention has been minimal compared to a new bulk water scheme, called the Hogenakkal Water Supply and Fluorosis Mitigation Project, that now supplies water to over 3 million people across 7,000 villages in three districts of Tamil Nadu. The scheme draws water from the river Kaveri over 100 km to the villages in this case study. It delivers this to village reservoirs at a highly subsidised price to Gram Panchayats meaning that they now enjoy a secure supply. The TWAD Board leads the project but it has outsourced the construction and operation of the scheme to the private sector through 'BOTT' (Build, Operate, Train and Transfer) contracts. The work has been divided out into five work packages with Larsen and Toubro (L&T) taking on the contract as bulk provider in the Morappur area. As part of that contract, it directly supports an operator in each Gram Panchayat who is supported by a specialist technical support team operating at the block level. Distribution remains officially in the hands of the Gram Panchayat with this managed either directly by the Panchayat or through the sub-committee of the VWSC, but noting the limited role of VWSCs in Tamil Nadu.

Maharashtra Jeevan Pradhikaran and the 'utilitisation' of rural water supply

MJP is the mandated public body responsible for the provision of water supply and sanitation in Maharashtra. In the Purna river basin region, where Amravati district sits, there is saline intrusion of aquifers covering an area of nearly 5,000 km². Here, MJP provides surface water schemes to villages and as part of this strategy it developed the Shanoor dam project covering 156 villages and two towns with household piped water supply. The federal government provided 50 per cent of the initial financing cost, which was equalled by the state government, following a loan from the Housing and Urban Development Corporation. The MJP was then the implementing agency for the construction of the multi-village schemes and now also operates and maintains the system. As such, the MJP can be described as being both the enabling support environment and also the service provider. It support functions include monitoring system performance, water quality testing, water resources management and conflict management. MJP also controls service provision across the whole system, including the bulk water production and distribution management. Ten MJP engineers and nearly 200 other staff are required to manage the system, including time-keepers and valve-men to operate and maintain the distribution systems at the village level, as reported in the Chary Vedala et al. (2016c) case study report. In this sense the service provision has been completely professionalised within MJP and there is not direct community involvement in service provision.

However, VWSC are still established in each of the 156 villages as part of the scheme. In the case study report, Chary Vedala et al. (2016c) argue that the role of these institutions has moved away from service provision to a model closer to consumer councils within urban systems. However, rather than provide oversight of the MJP, which is a principle task of a consumer council, the core role of the

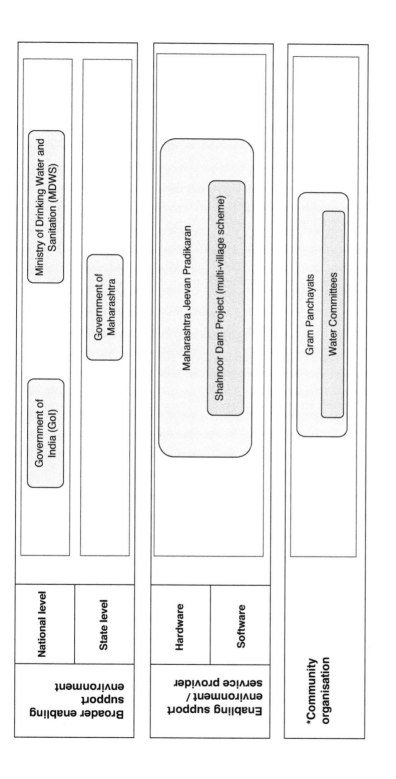

Figure 8.3 Simplified schematic of institutional arrangements for enabling support environment for Maharashtra case

Note: * the water committee does not play the role of service provider so this has been relabelled as 'community organisations' rather than 'community service provider'.

VWSC appears to be to promote compliance among the community for regular tariff payment. As all households have meters installed, the VWSC discourages and monitors for misuse and also plays a role mediating between MJP and any tariff defaulters. In addition to the VWSC bodies, the main involvement of the community in the scheme is through the payment of tariffs based on consumption of water, as measured by the meters. In this sense, the case reflects an example whereby rural water supply is moving closer to an urban-based model and the population move from being 'community members' to consumers. This is a possible trajectory of professionalising the rural water sector, a trend that has been identified as required if it is to move towards more sustainable water services (Lockwood and Smits, 2011; Moriarty et al., 2013). It is also a model that moves beyond the models of community management that were developed earlier in this research report towards this new category of 'urban-style utilitisation', a term which will be further clarified later in this chapter. The question whether this approach of switching to direct provision in the distribution system loses something compared to more active community involvement is engaged with in the discussion section of this chapter.

Coverage and service levels in the development cases

With the cases in this chapter all coming from the state-run programmes, and noting the positive correlation between state income levels and coverage levels identified in Chapter 4, it is unsurprising that the coverage levels from these case studies are some of the highest in the study. In all cases the level of household water supply coverage is above 90 per cent in the programme villages. This is above the state-wide averages (Census of India, 2011b), indicating relative success. From a policy perspective the villages are also above the revised NRDWP (Government of India, 2013a) 80 per cent target for household connections in rural areas which indicates that community management can play a role in supporting the required coverage rates to achieve this goal. Focusing on the service-level data, this is the only chapter in which the programme villages from each case presented meet the Government of India norms for acceptable service characteristics, as per the consolidated service-level indicator. The reason for this is because of the high

Table 8.2 Service levels in the development states

Case no.	State	Service level (median)	Percentage of population reaching basic or above service level
15	Gujarat – WASMO Gandhinagar	Improved	87%
16	Gujarat – WASMO Kutch	High	98%
17	Tamil Nadu – Local self-government	Improved	63%
18	Tamil Nadu – Public-Private Hybrid	Improved	53%
19	Maharashtra	Improved	94%

proportion of household connections found with this type of technology delivering significantly higher service levels than any other water point (the evidence for this is expanded on in Chapter 10).

Costing community management in the developmental states

There is significant diversity in the reported capital expenditure from village to village. The highest cost from the whole research is the $800 per person capital investment for a multi-village scheme in Maharashtra. This is due to the significant capital costs associated with developing the Shanoor dam and multi-village scheme, which make it over five times higher than any other village, demonstrating the additional investment required for this type of sophisticated multi-village scheme. The Maharashtra case should, however, be considered in comparison to the Tamil Nadu Morappur case. The capital expenditure that is displayed in Table 8.3 is for a single-village scheme made in 2005, which was then augmented into the Hogenakkal multi-village scheme in 2012. Yet the cost of the Hogenakkal investment has been included in the on-going operational costs of the scheme rather than the CapEx costs here. This is due to the financing mechanisms which underpin each scheme. In Maharashtra, the Government of India and the state Government of Maharashtra, financed the project through a model that does not cover the depreciation costs of these assets, as per a fixed-assets accounting model. This effectively means that government has made a one-off lump-sum capital investment with no financially linked plans to recuperate these costs for subsequent capital investment at the end of the infrastructures life.

In contrast, in the more recent Hogenakkal bulk water scheme, the Government of India and the state Government of Tamil Nadu have factored in depreciation cost meaning the financing of the depreciation of the asset is paid for over the life of the infrastructure by the state agencies responsible for running the service. The

Table 8.3 Financial resource implications of community management in the developmental states

Case no.	State	Capital expenditure (CapEx)	Percentage support contribution to CapEx	Annual recurrent costs	Percentage support contribution to recurrent costs
15	Gujarat – WASMO Gandhinagar	$73	92%	$6	73%
16	Gujarat – WASMO Kutch	$196	91%	$9	52%
17	Tamil Nadu – local self-government	$128	90%	$28	75%
18	Tamil Nadu – public–private hybrid	$17	91%	$46	72%
19	Maharashtra	$1,019	100%	$13	53%

costs here therefore become recurrent costs as per this research's costing framework tool. The difference in this financing models is thought to partly reflect a trend towards more sophisticated and sustainable financing models in India as it shifts from the expansion phase to sustainability phase in rural water supply (and infrastructure development more generally). By integrating the depreciation costs into the operational costs of the scheme, the government is following a more sustainable model that means it should have the required capital for next generation capital investments at the end of the infrastructure's lifetime. Given the financing arrangements described before it is unsurprising that the Tamil Nadu public–private hybrid case study has the highest recurrent costs, including the bulk water cost. The lowest recurrent covers are from the highly participatory Gujarati schemes.

Discussion of the developmental cases

The discussion section now develops two themes from this chapter that provide insights into the future trajectories of development of the community management model (and rural water supply more generally) in India. It first expands the discussion and analysis of community management in multi-village schemes and second examines institutional change in PHEDs.

The limits of subsidiarity with technical sophistication

The Government of India has specified that 'rural domestic water supply should preferably be from surface water' (Government of India, 2013b). Looking at the broad sweep of post-independence history in India, this represents a significant shift from the 1970s and the 'million wells' borehole drilling programme of that period (James, 2004). The switch to surface water means the technical design of rural water supply schemes is becoming more sophisticated, involving water treatment processes and large piped networks, and a general trend towards multi-village schemes. As considered in a World Bank discussion paper on the topic of multi-village schemes in India (World Bank, 2001b), the approach both has potential for driving investment efficiency through economies of scale and also for raising overall service levels and professional standards yet, equally, there is significant danger that poorly implemented multi-village schemes lead to diseconomies of scale as the size and complexity of the system leads to unforeseen problems and worse outcomes than through alternative models. Moreover, they represent a particular challenge for how best to maintain the principles of subsidiarity which underpin the constitutional mandate for village-level management of water supply, when much of the management challenge in multi-village schemes is beyond a single village. Considering the potential complexities and variety of models for achieving this it is surprising that the NRDWP rarely mentions multi-village schemes. The key references are given here:

> All water supply schemes within the GP shall be maintained by the Gram Panchayat. For multi-village or bulk water supply schemes the source, treatment

plants, rising mains etc., shall be maintained by PHED or the concerned agency while the distribution and other components within the village are to be maintained by the GP. State Governments shall endeavour to develop sustainable sources of funding for maintenance of rural water supply schemes and shall ensure that the Central and State Finance Commission and O&M funds released by MDWS are released to Panchayats.

(Government of India, 2013a)

For multi-village schemes, the standing committee of the Block Panchayat could perform a similar role [as the VWSC in single-village schemes].

(Government of India, 2013a)

While in some ways the policy portrays the division of responsibilities between the Gram Panchayat and higher bodies as simple and definite, there are clearly numerous configurations that can be developed (for example, for a review of the kind of contracting models that could be followed see Franceys and Hutchings, 2017). The case studies in this chapter provide windows into two such approaches. In Maharashtra an approach that has been labelled as an urban-style utilisation model has been followed that does not follow the principles of the NRDWP as the state agency also takes on the role of service provider. Reflecting on the upper right of Figure 8.4 below, this is a model that has moved completely away from community

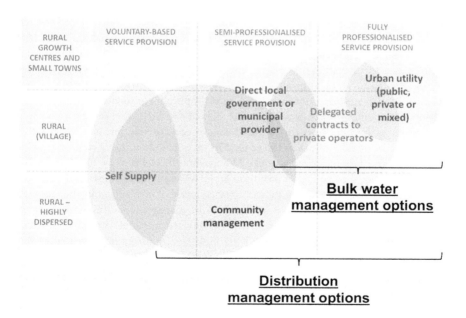

Figure 8.4 Rural water supply management models

Note: adapted from Lockwood and Smits (2011).

management towards 'utilitisation'. In Tamil Nadu, however, there has been the split between the bulk water and local distribution with a private BOTT contract for the bulk water scheme and then the Gram Panchayat taking on the (partial) community management of distribution.[1] This type of segmentation is analogous to what can be seen in other sectors, such as the split between transmission and distribution networks in the UK electricity sector (Ofgem, 2016). A key question for Indian policy-makers, at state and federal level, is whether a single or segmented management model will deliver better results. It is likely that this will depend on context and, while it is acknowledged that this research cannot answer that question conclusively, it can provide some valuable insights, particularly in terms of the role (or not) that community management can play within this set-up.

It will be widely accepted that bulk water production for regional multi-village schemes can only be managed by the fully professionalised models on the upper right-hand side of Figure 8.4. The critical decision is then how distribution is managed within the scheme, where there are various options that can include community participation or not. Looking at both the relevant cases from this chapter, there are lower levels of participation in capital expenditure and service provision, compared to the other cases in this chapter. Capital expenditure is rated as 'passive participation' compared to the 'interactive participation' seen in the single-village schemes from Gujarat and Tamil Nadu. This is simply because the capital expenditure has to be undertaken at a higher level and although there may be some forms of consultation with communities, the power that individual villages have to effect design and other factors is extremely limited in a regional multi-village scheme. In this sense, the study suggests that a 'high' level of participation in capital expenditure would no longer be feasible in regional-scale schemes. For the service delivery phase there are differences with Tamil Nadu being categorised as a form of functional participation, reflecting the executive role of the Gram Panchayat and semi-autonomous VWSC in distribution management, whereas in Maharashtra the role of the VWSC is purely consultative and the distribution management remains within the hands of the MJP.

The study does not have the required data to conclusively prove whether either approach is more effective than the other. However based on interpretation of what data is available it is suggested that the Maharashtra model is more likely to lead to problems in terms of equity of provision across villages. In that case study the tail-end village receives among the poorest service levels of any village in the study. As the VWSCs do not play a role in distribution management, this is thought to disempower the village from effectively holding to account the MJP provider. More generally, it points to the need for high-level management structures that have representatives from every community who can hold the service provider to account. The case discussed in the previous chapter from Kerala provides one approach for doing this through the formation of scheme-level executive committees (SLECs) that bring together representatives from individual VWSCs at the distribution level to take on the management of the entire scheme. It is noted that the size of that scheme is significantly smaller than the ones from this case study. However, the formation of higher-level community representative groups to hold the service provider to account would still be possible and it is thought that in Maharashtra the lack of community participation at the village level would impact the opportunity to deliver this.

Institutional reform for community management

In an earlier chapter of this report, focusing on the low-income states, the argument was put forward that to successfully support community management at scale then the most appropriate forms of support should involve both 'mechanical' organisational types as well as 'organic' organisation types (Burns and Stalker, 1961). This chapter continues to support this argument but showing that the WASMO support organisation has the best outcomes in terms of community participation and leveraging high levels of community tariff contribution, relative to the recurrent costs of supply. This has been achieved through its integrated support model, bringing together both software and hardware support to communities, which is reflected in the strong all-around performance on the institutional assessment tool, presented in Figure 8.5. MJP and TWAD Board retain a pattern of institutional performance that is reflective of the general PHED model with strong technical

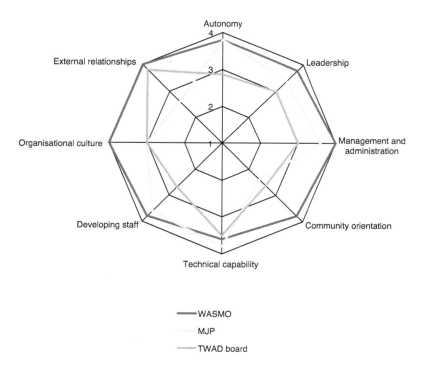

Figure 8.5 Subjective institutional assessment of public support agencies in developmental states (1 = low capability; 4 = high capability)

Table 8.4 Organisational data for developmental states

Case no.	State	Organisational characteristics (summary indicators)		Partnering typology	Participation in service delivery
		Professionalisation (out of 100)			
		ESE	CSP		
15	Gujarat – WASMO Gandhinagar	90	69	Operational	Interactive participation
16	Gujarat – WASMO Kutch	90	75	Operational	Interactive participation
17	Tamil Nadu – local self-government	45	81	Collaborative	Self-mobilisation
18	Tamil Nadu – public–private hybrid	75	69	Transactional	Passive participation
19	Maharashtra	55	33	Transactional	Passive participation

performance but low community orientation. There is nothing wrong with the more mechanical approach of the TWAD Board and MJP if they recognise the need to work with partners to deliver more participatory rural water supply, such as the TWAD Board's partnership with the PRIs, or if they follow a model not reliant on high levels of community participation as is the case in Maharashtra. Regardless of these points, the core argument that organic departments within PHED that can work closely with communities in a more creative, less-standardised fashion is conducive with high levels of community participation remains valid. This chapter, however, provides further insight into the processes of reform that can lead to these types of organisations. In particular, WASMO is a public organisation that has emerged from a process of institutional reform to support community management while TWAD Board represents a public organisation that has undergone a process of institutional reform for a similar purpose but which failed to materialise with such positive results.

The section now tells the story of two processes of institutional reform. It starts by focusing on the reform process in the TWAD Board that occurred from the early 2000s onwards. It started in 2002 when the TWAD Board was faced with four related challenges: responding to a 'water crisis' in which groundwater tables were falling; the exclusion of vulnerable groups for water services; external pressure to shift to a more demand-responsive approach to service delivery; and, institutional inertia due to powerful 'water technocrats' (Nayar and James, 2010). To respond to these challenges a Change Management Group was formed by a small group of engineers within TWAD Board that aimed to drive a reform process to:

- shift from an access approach to a service delivery approach;
- shift from providers to partners with the community;
- shift to a sustainable enhancement approach; and
- aim towards total institutional transformation.

In a number of workshops the Change Management Group summarised these aims into a working approach for engineers which was drafted into a statement that became known as the Malarimalai Nagar Declaration. It read:

> We will evaluate the existing schemes and ensure that the schemes are put into optimal use first. Then the rehabilitation will be undertaken wherever necessary along with revival of traditional sources. This will be taken up before taking up any new schemes in the block. We will aim at 10 per cent increase in coverage with the same budget.
>
> (Nayar, 2006)

The Change Management Group then set about getting their fellow employees to commit to the statement. Initially they hit resistance as the declaration was deemed to provide criticism of past practice, but the group managed to convince a secretary from the federal Ministry of Water Supply and Sanitation who was visiting Chennai to sign it. With this high-level support, employees throughout TWAD Board were keen to sign up. When the petition was passed to the TWAD Board senior management, they considered it in a board meeting but came to the conclusion that if engineers adopted the working practice it would threaten the long-term financial security of the organisation. This was because a large part of the organisation's operating budget came from overhead charges on capital expenditure investments into new schemes. As the new working approach advocated a bigger focus on the maintenance of existing infrastructure, this would reduce the need for capital expenditure therefore reducing the budget for the organisation. It was, in short, an example of the often seen perverse financial incentive that drives a bias towards capital expenditure within the water sector.

Undeterred, two committed members of the Change Management Group who were executive engineers at the district level decided to take a stand. They insisted as they had signed the declaration they were not going to identify new schemes in the next annual investment plan for their districts unless they were technically needed. This meant that across the two districts they submitted plans for only 18 new schemes when normally there would be 115 in each district. This led to a bitter stand-off in which senior management threatened to close down the district-level office and the executive engineers were faced with criticism from both above and below, as their junior staff feared losing their jobs. However, following clarification from Delhi over a secure financial settlement for TWAD Board, the senior management team accepted piloting the reform principles in a few projects. They commissioned what become known as the Change Management Pilot Programmes across 145 villages in the state. The official name of the programme was Tamil Nadu Rural Water Supply Programme (TNRWSP) and it was this programme that occurred in the Morappur and Erode case studies presented in this chapter. An impact evaluation of the pilot provided evidence of significant improvements in outcomes from the same budgets, including evidence of functioning VWSCs in every village and a reduction in OpEx costs of between 10–30 per cent across these villages (Nayar and James, 2010). It was at this stage that the institutional reform process was presented as a successful case study in reports for DFID, UN-Habitat, UNICEF and the Government of India.

However, the pilot projects were never fully scaled up into the wider working approach of the TWAD Board and, as reported in key informant interviews conducted for this research, many of the officials involved in the Change Management Group as well as the senior staff that were 'won over' were transferred out of the organisation. TWAD Board has remained a largely technically-focused – mechanical – support organisation, preferring to outsource the software element to other agencies or bodies, rather than take it on itself. It should be noted, however, that TWAD Board remains a high-performing PHED that delivers high coverage rates compared to many PHEDs in India. Yet the story of institutional reform in TWAD Board reveals that, even in this high-capacity state with an extremely well-planned and implemented reform process, delivering institutional change that transforms PHED-type organisations into what have been described as RRWSA is extremely challenging. As North (1990) points out in his work, processes of institutional change 'tend to produce a new equilibrium that is far less revolutionary [than intended]', which seems to be the case for the TWAD Board experiments to move towards a more community-management approach.

The change process in WASMO followed a different trajectory of development, although its origins emerged in a similar context, with the Gujarat Water and Sewerage Board facing a 'water crisis' and pressure to reform its largely supply-driven model to the more internationally favoured demand-responsive approach (Das, 2014). The reforms can be traced back to the Ghogha Rural Drinking Water Supply Project in 1997 in 82 villages from Bhavnagar district. Here, the Government of Gujarat with support from the Royal Netherlands Embassy funded a programme based on the principles of the demand-responsive approach to community management. The success of this programme coincided with the federally driven ideas around community management that were being trailed in the Sector Reform Projects from 1999. With encouragement from the success of the Ghogha project and favourable policy environment, the Government of Gujarat decided to establish a new body to support the community management of rural water supply, which was to be named WASMO, rather than try to reform the Gujarat Water and Sewerage Board to serve this purpose. This also meant a realignment of the Gujarat Water and Sewerage Board to become more focused on urban services and only play a role in the development of large-scale and complex rural water supply infrastructure, such as multi-village schemes.

Interviews with key informants on this matter during the course of the research suggested that this approach of establishing a new body rather than reforming the existing institution was only possible due to the strong political support received from the Chief Minister of the state, Nerenda Modi. As James (2011, p. 56) explains, when assessing change in the Indian rural water supply sector:

> Political support is vital, especially to insulate reform processes from vested political interests ... For instance, in Gujarat, the support for the initiative from the Chief Minister of the state ensured that local politicians did not try to manipulate the scheme for personal political gains.

Moreover, WASMO was established therefore as a Special Purpose Vehicle with independence from the Gujarat Water and Sewerage Board, allowing it to use government funds but have a reasonable degree of autonomy. Such parastatal bodies are reported to have a strong tradition in Gujarat compared to other states, which may have contributed to the acceptance of the model (Das, 2014).

Either way, these factors combined to enable the establishment of WASMO but it was the role the organisation played in the long-term response to the Gujarati earthquake that helped build its legitimacy and expand its influence throughout the state (Das, 2014). The earthquake struck in 2001 killing 18,602 people and injuring over 166,836 (Gujarat State Disaster Management Authority, 2003) but it was in 2003, with many communities still struggling to recover from the earthquake, that WASMO started the Earthquake Rehabilitation and Reconstruction Project in Kutch. The project aimed to provide assistance to villages in developing water supply and sanitation systems that had been affected by the disaster. Using the additional money that was available through public reconstruction budgets from the federal government it was able to demonstrate it could successfully work at scale. At the same time, its role in the state was further boosted by the convergence between its approach and the federally driven Swajaldhara programme to scale-up community management. By 2004 the organisation no longer received funding from the Dutch Embassy or other international supporters and was operating a community management programme across Gujarat based completely on domestic resources (Das, 2014). As such for over the past ten years it has been a rare example of a domestically driven and domestically financed highly participatory community management model in India. However, while WASMO has been identified as a flagship example for a number of years, no other state government has been able to follow the model as successfully.

The lessons from the WASMO and TWAD Board reform process is that reform processes are long, on-going and non-linear. In India reforming large mechanical public bureaucracies is a particularly testing form of institutional reform. The two experiences presented here suggest that it may be more appropriate to create new institutions or split out functions into new independent organisations to take on organic approaches, such as WASMO's. Rather than try to deliver organisational change to shift a largely mechanical organisation towards more organic working practices (which may not necessarily be a good approach anyway as mechanical working approaches are more efficient at scale and for standardised processes), which appears to be the story of TWAD Board, more broadly, the WASMO approach can be described as being more top-down with high-level political support being critical to pushing through the reform. In contrast, the TWAD Board story is in part an empowering story of a bottom-up movement. Yet in the end it achieved far less revolutionary change than it had intended. It is contended that top-down change in mechanical organisations is more likely to succeed and probably even more so in states that have developmental tendencies that lean towards a preference for the centralisation of power.

Conclusion

As India's states advance along the economic transition curve, this chapter showed how the conventional community management model for rural water supply is being challenged by a move to more technically sophisticated multi-village schemes (MVSs). It examined this trend by presenting three cases from the richer states of Maharashtra, Tamil Nadu and Gujarat reflecting different intensities of what may be called a 'utilitisation' of rural water supply. This process of 'utilitisation' reflects a shift in which an increasing level of control is exercised by government agencies over water supply, bringing with it higher levels of system resilience but potentially additional costs and lower levels of community participation. The chapter also discussed two examples of domestic reform found in India. These demonstrated the value of high-level political support to the long-term success of reform within the Indian context.

Note

1   Similarly, in Gujarat, under WASMO, there are villages supplied through bulk water but retain a community management element – with the WASMO villages likely to follow an even more highly participatory community local distribution management model. However, these were not captured in this study so cannot be extensively commented on during this discussion section.

9 Community management in the mountains and hilly regions of India

Unlike the other empirical case study chapters, this one presents the data from case studies that share similar geographical contexts but have varying socio-economic statuses. All the cases presented in this chapter are from either largely mountainous states, situated in the Himalayan region of India, or the hilly northeast region located away from the rest of India in the area in between Bangladesh, Bhutan, Myanmar and China. In Chapter 4 it was shown that while most Indian states were reaching universal levels of improved water supply, the mountainous and hilly states were lagging in terms of basic improved access. As such, it was argued that the physical geography of such states means it is useful to treat them as a separate class, as the remoteness of the mountains means that a greater proportion of the population live without basic access. Yet, as also explained, when looking at the piped water supply access rates, such states are above the Indian trend line largely due to the prominence of gravity-fed piped systems and the difficulty of installing hand-pumps. This is the paradox of the mountainous states – they are below the trend line for basic improved supply but above the trend line for access to piped water supply. In this chapter, the case studies come from the states of Meghalaya, Himachel Pradesh, Uttarakhand and Sikkim, which are states that go right across the economic development spectrum in India. They are the 15th, 10th, 8th and 1st richest from this study, respectively.

In terms of enabling support environment set-up, the cases can be classified across three spectrums. In the poorest state, Meghalaya, and also the richest, Sikkim, the cases can be described as being examples of the local self-government management system found in India. Intensive support is provided in the implementation phase by specialist centralised agencies but then the local self-government institutions take on service delivery tasks with limited support from the centralised agencies. In Himachal Pradesh the case study focuses on villages that were part of a public–donor partnership pilot project run by the Irrigation and Public Health Department, Government of Himachal Pradesh, with support from the bilateral donor Deutsche Gesellschaft fur Internationale Zusammenarbeit (GIZ). However, now the pilot project has finished, communities are left to manage services in relative independence and so this case study can be considered a form of community management without direct support. The final case study in this chapter is from Uttarakhand and is focused on the civil society model. Here, a national funding

Table 9.1 Overview of the enabling support environment and community service providers

Case no. (inverse GDP rank)	Case name	State	Enabling support environment	Community service provider/ community organisations	Scale of support programme (approximate population covered)
5	The Dorbars and gravity-based piped water supply in Meghalaya (Saraswathy, 2016b)	Meghalaya	PHED and Social and Conservation Department, Government of Meghalaya	Dorbor Water Supply and Sanitation Sub-Committee	2.4 million (rural population of state as reported in census 2011)
10	Community Water Classic: the success of community-managed water supplies in Himachal Pradesh with limited on-going support (Harris et al., 2016c)	Himachal Pradesh	Rural Management and Development Department and the Panchayat Raj Department, Government of Sikkim	Village Water and Sanitation Committee as sub-committee of the Gram Panchayat	456,999 (rural population of state as reported in census 2011)
12	Support to community-managed rural water supplies in the Uttarakhand Himalayas – the Himmotthan Water Supply and Sanitation initiative (Smits et al., 2016)	Uttarakhand	Irrigation and Public Health Department, Government of Himachal Pradesh and GTZ (Deutsche Gesellschaft für Internationale Zusammenarbeit)	Village Water and Sanitation Committee (unregistered)	2,500 (estimated population across the nine pilot villages)
20	Decentralised local self-government and gravity-based piped water supply in Sikkim (Saraswathy, 2016c)	Sikkim	Himmothan Society and the Himalaya Institute and Hospital Trust	Village Empower-ment Committee (registered under the Societies Act)	2,603 population supported by the Himalaya Institute and Hospital Trust

body called the Sir Ratan Tata Trust has set up a specialist NGO to support rural water supply in the state of Uttarakhand called the Himmothan Society. The Society funds and administers its work through local NGOs, with the one studied in this case study called the Himalayan Institute and Hospital Trust (HIHT) which undertakes the direct work in villages. Village Empowerment Committees act as the service provider based on a model of community management with direct support.

Local self-government cases – Meghalaya and Sikkim

This section focuses on the case study from Meghalaya and Sikkim. Meghalaya is known as the 'wet state' with an average annual rainfall of 2.8 metres which rises to 12 metres in the wettest zone. Alongside its north–eastern neighbours of Assam, Tripura and Mizoram, much of the state is governed under the Sixth Schedule of the Indian constitution that gives additional autonomous power to areas dominated by Schedule Tribes (Constitution of India, 1950). This has led to the development of a distinct institutional structure of local government compared to the Panchayat Raj system found in the rest of India. In areas governed by the Sixth Schedule traditional tribal councils have been empowered to become part of the government systems. In East and West Khasi Hills, where this case study is located, this system is known as the Dorbar system. The Dorbars share a similar three–tied structure as the Panchayat Raj Institutions, with the lowest level of the Dorbars sharing the same administrative scale as a Gram Panchayat. However, they also have unique features such as an elected membership consisting solely of men and greater power regarding subjects such as law and order. Similarly to the Gram Panchayat, they take responsibility for public service delivery within villages including rural water services. In this sense, the case from Meghalaya is similar to the Gram Panchayat-managed supply reported on in the case studies from Chhattisgarh and Tamil Nadu, in that, the distinction between community management and highly decentralised public provision is conceptually blurred.

In Meghalaya there are two key agencies in the enabling support environment although almost all activities they undertake are in the implementation stage only (or when major capital works are required during rehabilitation or expansion). This includes the PHED that develops gravity-fed piped water supply schemes across the state. In Meghalaya the PHED focuses on this task and then passes the infrastructure over to the Dorbars for service delivery, providing neither software support during implementation nor any on-going support apart from in emergency situations. Similarly, the Soil and Water Conservation Department develops what it calls 'spring tapping chambers' in or close to villages which become alternative but popular communal water sources which are particularly popular as a source for drinking and cooking (Saraswathy, 2016b). These spring tapping chambers are then passed to the Dorbar for maintenance and the department provides no on-going support. This means that the villages covered in this case have dual systems, with piped water supply complemented by spring sources. The villages through the Dorbars operate and maintain these dual systems in an independent fashion with no or very limited support from either of the centralised agencies.

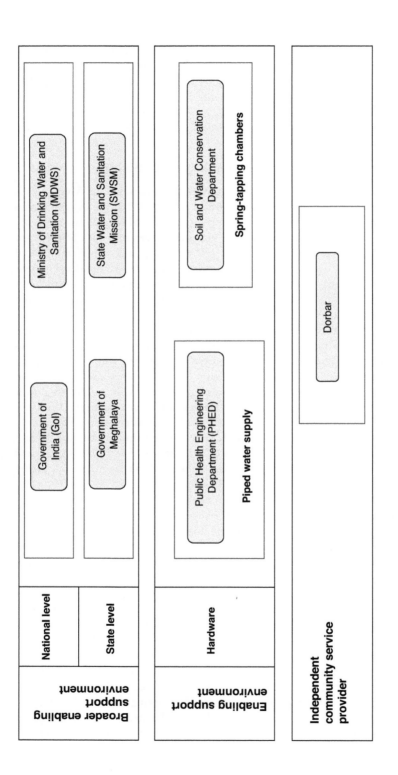

Figure 9.1 Schematic of institutional set-up in Meghalaya case study

The technological set-up is considered to play a role in this as the villages are served by gravity-fed piped water supply systems, which are expensive and complex to construct, but which are generally easier to operate as they do not have motorised pumps or high energy costs (as discussed in more detail in the costing section later in this chapter). The strength of the Dorbars mean that they are able to deliver under this model, with this especially apparent in Kheuih Shnong Shora village which has been managing a gravity-fed piped system that was first developed around 1965 (ibid.).

Sikkim on the other hand is the richest state from which a case study has been selected with a GDP per capita of over $10,000 (PPP) (Reserve Bank of India, 2015). By Indian standards it is an extremely small state with only 620,000 people with the 31st lowest population density out of 35 states or union territories (Census of India, 2011a). Home to the third highest mountain in the world, it has an altitude of between 300 and 8,500 metres, meaning that snow melt and glaciers play a crucial role in water management. The Rural Management and Development Department is the public agency mandated to deliver rural water supply across the state and it does this through largely implementation-driven activities in which it develops infrastructure to pass onto the Gram Panchayats for on-going operation and maintenance. However, in contrast to Meghalaya, it places an emphasis on providing training to Gram Panchayats through its State Institute for Rural Development that delivers training to the local bodies to support the management of water supply and other public services. During the implementation stage it embeds a junior engineer within the Gram Panchayat to oversee any works and to train what it calls 'barefoot engineers' within the village, which is the equivalent of the waterman or pump operator in other states, who can deal with any minor problems with supply going forward.

Although it is rated as 'only' ninth in the Government Devolution Index, Sikkim is widely recognised as having been an early adopter of decentralisation, having passed the Sikkim Panchayat Act in 1993 immediately following the 73rd and 74th constitutional amendment. This decentralisation means that the Gram Panchayats in Sikkim has over 20 years' experience in developing annual development action plans. Over this time it has become common for cooperative societies to develop within villages to execute public work contracts and local governments have the power to leverage different funding sources towards delivering on development plans. For example, the MGNREGS (rural employment guarantee scheme) is commonly used to fund water security activities such as building rainwater harvesting structures. The villages studied come from three of the four districts in Sikkim as best practices are common across the state and the field research teams were keen to reflect this in their village selection. In all cases there is a gravity-fed single-village scheme developed from a spring source that provides household connections throughout the village. The Gram Panchayats take on service delivery with autonomy from other agencies but they receive a INR 100,000 block-grant each year to cover these costs, which it supplements through a tariff charge of INR 20 per month from each household. Overall the Sikkim represents another example of extremely advanced government decentralisation to the Gram

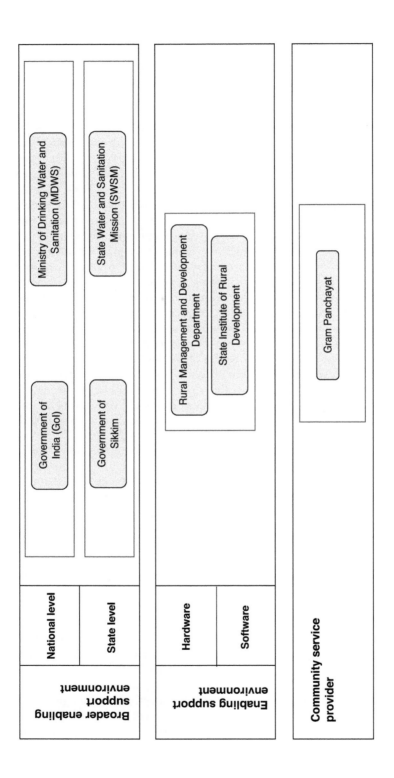

Figure 9.2 Schematic of institutional set-up in Sikkim case study

Panchayat rather than autonomous community management as understood in the international literature.

Forgotten communities in Himachal Pradesh

The case study from Himachal Pradesh focuses on a donor-supported pilot project that was implemented in nine villages between 2005 and 2011. GIZ, the German agency, worked with the Irrigation and Public Health Department, Government of Himachal Pradesh, with a mandate to pilot and refine a community management model that could be scaled up across the state. This was undertaken during the period of Swajaldhara and pressure from the federal government for states to move towards community management. In the end the pilot model trailed was not taken up by the Irrigation and Public Health Department, yet the legacy of the pilots provides an example of the type of community management that can work in isolation. That is, an approach with intensive support provided to communities during the implementation of infrastructure then a withdrawal of support organisations, in this case the literal end of the pilot and therefore the closing of the project unit within the Irrigation and Public Health Department, and communities left to manage their water services in relative isolation through VWSCs. This sequence of events means the villages are now the responsibility of the standard Irrigation and Public Health Department support programme, but the engineers here are reportedly hesitant to help these villages as they believe they are not part of the main government programme (Harris et al., 2016c). This is one of the dangers over the long term with pilot projects and fragmented programmes as the primary government agency in many cases will ultimately have to take back responsibility of the scheme. In Himachal Pradesh this causes problems as the Irrigation and Public Health Department follows a much lower intensity of community management, with communities only expected to take on the operation of schemes, but not formally own them nor be involved in maintenance, which remains the responsibility of the Irrigation and Public Health Department (ibid.).

The size of the villages covered in this case study are considerably smaller than the pan-Indian average of around 5,000 people, with each having between 200 and 400 habitants. In each village a VWSC takes on the operation and maintenance of supply through a voluntary approach with what were described as 'informal' service provision arrangements, with limited record keeping (Harris et al., 2016c). With the type of gravity-fed systems developed, which required no on-going energy costs, and such small villages, it is contended that this type of rudimentary community management can play a role in service delivery in such contexts. In this sense, the case study is reflective of what is found in many Sub-Saharan African countries where infrastructure is completely handed over to villages in sparsely populated regions (Harvey and Reed, 2006), although the system here is still piped water supply rather than the more common handpumps found in many African countries. However, even within the three pilot villages studied in this case study, the problems that have been reported with isolated community management had already emerged in Paddar village:

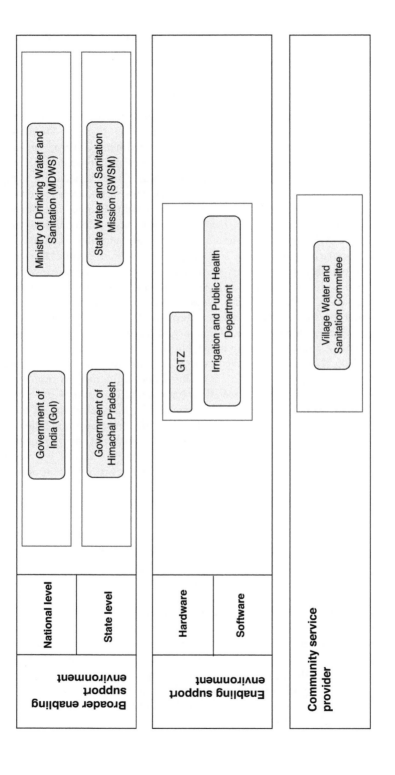

Figure 9.3 Schematic of institutional set-up in Himachal Pradesh case study

People did not feel capable of maintaining the system, with a sense of being overwhelmed by the responsibility. There were numerous calls for the IPH to take over the system (perhaps prompted by the research team being accompanied by IPH staff). It was also felt that there was no point in charging a tariff as the amount collected would not be sufficient to employ anybody to maintain the system.

(Harris et al., 2016c, p. 22)

The community here were left isolated and were keen to move to the lower-intensity public-community managed system offered by the IPE. Ironically, such a set-up was taken at the control village in Pali where community members explained that it would be 'impossible' to manage the system themselves and they were glad that it was the responsibility of the IPE (Harris et al., 2016c). In this sense, this case study provides a warning about the limitations of community management more than a success story.

Civil society in Uttarakhand

Uttarakhand is another state where a large World Bank supported community management rural water supply programme can be found. That programme is officially called the Uttarakhand Rural Water Supply and Sanitation Project but is commonly known as Swajal (Smits et al., 2016). Yet with a number of World Bank programmes selected from other states, it was deemed appropriate to focus on a civil society model that operates in the state. In this case the Sir Ratan Tata Trust funds an NGO called the Himmotthan Society to support development activities in the villages of this Himalayan state. That society in turn manages a programme that implements and supports rural water services to over 632 villages with an emphasis on serving small and remote communities (the average village size being just 43 households). It does this in coordination with government having its plans approved by the State Water and Sanitation Mission each year and having consolidated this coordination by signing a Memorandum of Understanding (MOU) with the state government in 2014 (ibid.).

As indicated in Figure 9.4, the Himmotthan Society provides project management but it works through other institutions to directly deliver support. This includes, in this case study, the Himalaya Institute and Hospital Trust, which is a local NGO that works within villages, and also the ENV DAS private contractor that supports the design and construction of infrastructure. Under this enabling support environment, the institutional system works by establishing a Village Empowerment Committee within habitations to take on the operation and maintenance of water services. These committees are now officially sub-committees of the Gram Panchayat following the signing of the MOU between Himmotthan and the Government in 2014.

The villages studied were small, concentrated Himalayan settlements in Tehri Gahrnal district. Historically each village was served by either mountain springs or a single public stand-post from an ageing government system (Smits et al., 2016).

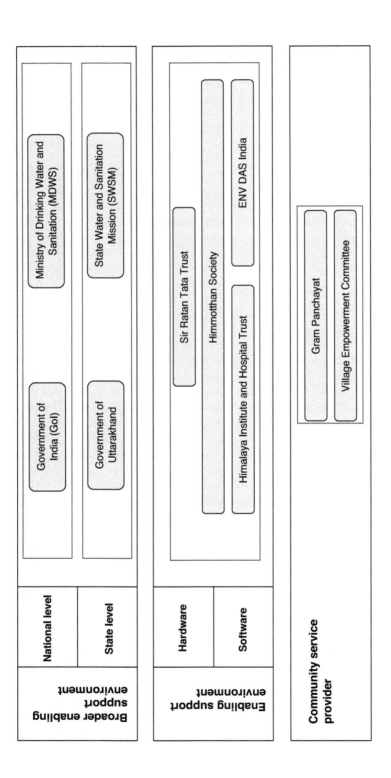

Figure 9.4 Institutional set-up of the Uttarakhand case study

Therefore, these villages had approached the Swajal office to be included in the government programme, but as this was not possible their cases were passed to Himmotthan to take on. In each programme village a gravity-fed piped system was developed and the Village Empowerment Committee established. The projects run through a four-stage cycle that includes feasibility assessment, planning, implementation and post-construction support. Unlike many other cases, the communities offer three technical choices – including a gravity-fed piped water supply, pumped piped water supply and rainwater harvesting structure – during the planning stage although it is unclear to what extent this choice is shaped by the feasibility of different designs, as invariably gravity-fed systems are reportedly developed (ibid.). A key element of the post-construction support is to set up insurance of the scheme from natural disasters, which are common in these mountainous areas. The insurance mechanism here has proven to be extremely successful in that it has helped the villages recover from the 2013 floods that wiped out many pipe networks in the area. Through such initiatives and the close relationship between the NGOs and government, this case study shows a sophisticated civil society model for supporting community management in a challenging mountainous context.

Service levels in the mountainous states

As with all the cases included in this study, all households in programme villages report having access to an improved water source. The surveyed population in Meghalaya reported having a dual-supply system with households using a protected spring for drinking water and cooking but also the piped supply system, either through household connections or public stand-posts. In Himachal Pradesh the surveying was able to cover over half the households in the programme villages but as these were so small, the sample was limited to 60 households across the programme villages. Nearly all of these had household connections from the gravity-fed piped system developed by the IPE and GIZ. Uttarakhand and Sikkim both have gravity-fed piped water supply systems in the programme villages but in Sikkim 100 per cent of households surveyed reported having household connections whereas in Uttarakhand all households reported having public stand-posts.

Table 9.2 Service levels in the mountainous states

Case no.	State	Service level (median)	Percentage of population reaching basic or above service level
5	Meghalaya	Sub-standard	45%
10	Himachal Pradesh	High	65%
12	Uttarakhand	No service	0%
20	Sikkim	Sub-standard	45%

The service levels are also variable across the cases and reflect the general pattern found in this study regarding the association between household connections and high levels of service. Himachal Pradesh is the only case where more than half of respondents reported accessing services that met all government norms, as per the service level indicator. In Uttarakhand – a case served exclusively by public stand-posts – all households receive either 'sub-standard' or 'no service' service levels. The service levels generally fail on both quantity and accessibility, with 84 per cent of respondents reporting using less than 40 lpcd, and 93 per cent of households reporting spending over 60 minutes a day collecting water. Meghalaya has the most mixed service levels across houses with over one-third receiving high service levels and over one-third reporting no service. The third that report no service levels are again public stand-post users who do not report using more than the 40 lpcd benchmark needed to achieve at least basic service levels. In all cases the control villages have comparable or worse service levels than the programme villages.

Costing community management in the mountainous states

In terms of financing, capital expenditure is above average in three of the case studies, which is to be explained by the nature of rural water schemes in mountainous contexts. In such areas gravity-fed schemes are used which are by their nature more technically sophisticated than groundwater-based schemes. The cost of construction is also generally higher due to the logistical challenges of working in hilly or mountainous contexts and the economies of scale are reduced when working in such small habitations. With the design of these schemes intended to avoid the need for pumping, the recurrent costs are often lower. This can be seen in the Meghalaya, Himachel Pradesh and Uttarakhand case studies that have very lower recurrent costs, compared to the other groups of case studies and as analysed in greater detail in Chapter 10.

Table 9.3 Financing of community management in mountainous case studies

Case no.	State	Capital expenditure (CapEx)	Percentage support contribution to CapEx	Annual recurrent costs	Percentage support contribution to recurrent costs
5	Meghalaya	$85	95%	$6	61%
10	Himachal Pradesh	$342	97%	$6	3%
12	Uttarakhand	$536	91%	$13	54%
20	Sikkim	$251	98%	$73	94%

Discussion of community management in the mountainous states

The further analysis and discussion in this chapter focuses on the role of community management in promoting efficient service provision in remote and challenging contexts. In reflecting on the shift to universality in the Sustainable Development Goals, a key challenge identified for the global sector is that progress in expanding access tends to slow as countries approach universal coverage (Fuller et al., 2016). This is partly because a significant proportion of the final 5–10 per cent of the unserved population tend to be found in remote settlements. It is such contexts that tend to challenge the efficiency frontier for formal service delivery options, with per person costs being comparatively high due to the diseconomies of scale of serving small and scattered communities. The case studies in this chapter, especially the two from Himachal Pradesh and Uttarakhand, help provide insights into the role community management may be able to play in this process. First, the evidence here confirms that there are significant additional capital costs during the implementation of schemes in small, mountainous villages. This is due to the complexities of developing gravity-fed systems in hilly terrain that require a minimum amount of fixed assets, even for these extremely small settlements. This is also compounded by the logistical difficulties of construction work in such contexts. The high capital costs are partly off-set by the lower, on average, operational costs of running gravity-fed systems as the cost of maintaining and powering motorised pumps is not needed.

The balance of high implementation costs and lower operational costs has led to some similar patterns of support across the case studies. As shown in Table 9.4 below, this is reflected in interactive participation scores from the participation scoring tool for all the case studies within the service delivery phase. In each case the community is devolved responsibility for service delivery, yet the way this is operationalised differs. In Sikkim and Meghalaya it is through the highly

Table 9.4 Organisational data on the mountainous case studies

Case no.	State	Organisational characteristics (summary indicators)			
		Professionalisation (out of 100)		*Partnering typology*	*Participation in service delivery*
		ESE	*CSP*		
5	Meghalaya	50	33	Transactional	Interactive participation
10	Himachal Pradesh	60	44	Collaborative	Interactive participation
12	Uttarakhand	65	56	Operational	Interactive participation
20	Sikkim	80	83	Operational	Interactive participation

decentralised local self-government of the Indian state while in Himachal Pradesh and Uttarakhand it is independent community institutions, registered under the Societies Act, which reflect the internationally recognised model of community management. In terms of the professionalisation of the service provision arrangements, the local self-government of Sikkim ranks highly while Meghalaya, also local self-government, and the Village Empowerment Committees of Uttarakhand are rated as medium in terms of professionalisation.

In Himachal Pradesh, however, the community service provider operates through a largely voluntary and ad hoc approach, scoring a ranking of no professionalisation in two of the programme villages. This is because there are no elections or formal nomination processes for those wishing to join the VWSCs, no systematic record keeping of income and expenditure, no systematic water quality testing and low general levels of professional practice. The VWSC has been left to manage its affairs without continued support and, whilst the service is still running eight years after implementation, the research found that much of the service provision was completed in an ad hoc and unprofessional manner (Harris et al., 2016c). The on-going sustainability of the system is therefore doubted without additional support.

Conclusion

This chapter has focused on community management within the mountainous and hilly states of India. Due to the remote nature of many villages within those regions, the natural preference for a community management model is perhaps even stronger than elsewhere in the country. However, the context does present challenges in terms of providing support to communities. The case study from Himachel Pradesh demonstrates the potential for communities to become isolated when the enabling support environment fails to operate properly, while the case study from Sikkim shows how effective support can be provided through the local self-government. This book now goes on to consider the overall trends across the case studies and different groupings.

Part III

Synthesis of successful community management arrangements in India

10 Organisational arrangements for successful community management

A key emphasis in this research is that there is a need to better understand how to structure support services to successfully enable the community management of rural water services (Baumann, 2006; Lockwood, 2002, 2004; Lockwood and Smits, 2011). This chapter is designed to answer that question by considering the *type* and *characteristics* of organisational arrangements from across the 20 case studies. Based on the basic conceptual framework for rural water services, the chapter reviews the findings at the enabling support environment and community service provider levels. It does this through presenting a series of typologies of organisations at each level before considering the characteristics that can be associated with each typology.

Types of enabling support environment

This section presents the analysis of the enabling support environment across the case studies. This led to the classification of four typologies of arrangements that are considered useful for distinguishing between different types of support systems. There are two forms of government support systems that are labelled as centralised 'state rural water supply agencies' (SRWSA) and decentralised 'local self-government' (LSG) support, which partly reflect the different ways decentralisation has played out across Indian states. Another typology is called the 'hybrid support approach', which involves public partnerships between government agencies and external agencies, such as donors, NGOs or the private sector. The fourth category is labelled as 'external agency support', which involves cases whereby NGOs or similar organisations (i.e. social enterprises) take on the role of an enabling support environment beyond the government system. Within these categories there are some sub-types of support systems while the distinction between one typology and the other is often not completely distinct. For example, the government support systems can sometimes outsource minor functions to non-government entities such as NGOs. However, as will be shown, these four typologies are considered conceptually distinct approaches that provide a useful set of categories to compare the cases by. Table 10.1 shows how the case studies have been allocated, but each typology is explained in detail in the following sections.

Table 10.1 Overview of case studies by type of enabling support environment

Centralised state rural water supply agency	Decentralised local self-government	Hybrid (public– donor/NGO/ private partnership)	External agency typology
1. Jharkhand	13. Kerala II	2. Madhya Pradesh	3. Odisha
4. Chhattisgarh	17. Tamil Nadu I	7. West Bengal	8. Telangana
5. Meghalaya	20. Sikkim	9. Karnataka	12. Uttarakhand
6. Rajasthan		11. Punjab	
15. Gujarat – Gandhinagar		14. Kerala I	
16. Gujarat – Kutch		10. Himachal Pradesh	
19. Maharashtra		18. Tamil Nadu II	

Centralised and decentralised government support

This section focuses on the two government-supported enabling support environ- ment typologies that include eleven of the case studies. It first focuses on the centralised SRWSAs, which it will be argued have an institutional set-up that has its legacy in the supply-driven approach that the Government of India took prior to the Sector Reform Projects of the late 1990s (James, 2011). During that period the primary concern of government agencies was concentrating on expanding access to rural water services and this approach involved investing power in a centralised agency that had the primary function of infrastructure asset creation (James, 2004). The body that undertook such work was conventionally called a public health engineering department or 'water board', although a variety of names were used. In the centralised SRWSA case studies, this hardware-focused agency continues to be the primary provider of support to community management. This is both in the implementation stage of infrastructure asset creation but also in the on-going support during the service delivery phase, as shown in Figure 10.1. The centralisation of responsibility for CapEx is a common approach across an interna- tional context but the centralisation of on-going OpEx support is less common (Lockwood and Smits, 2011). Yet in India the centralised SRWSA model is the most common government model reported on in this study although there are subtle differences between the seven case studies exhibiting this support.

They can be divided into two broad groups which include the 'standard' SRWSA found generally in the poorer states and what is described as a 'Reformed SRWSA' that is shown in the two Gujarat case studies. The 'standard' SRWSAs reflect the hallmarks of what has been described, in the context of Indian sanita- tion policy, as the 'technocratic governing machinery' of the Indian state which is 'a hierarchical and technocratic bureaucracy that is well suited to send down tech- nical designs and subsidies for physical infrastructure projects' (Hueso and Bell, 2013, p. 1013). As shown in Figure 10.1 and exhibited in the case studies from the poor states of Jharkhand (Javorszky et al., 2015), Chhattisgarh (ibid.) and Rajasthan (Harris et al., 2016a), the SRWSA develop infrastructure and then provide support

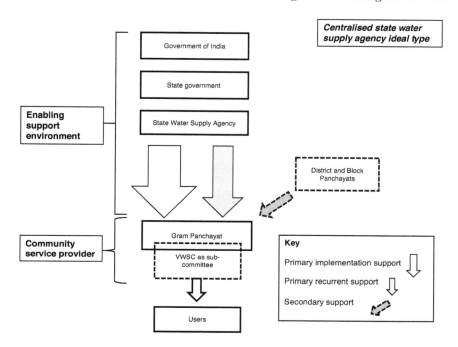

Figure 10.1 Institutional map of centralised state water supply agency enabling support environment

during this implementation period. For example, in Chhattisgarh, immediately following implementation the SRWSA operate the service for a transitional period of three to six months during which members of the community can shadow the agency's staff and learn how to operate the water supply (Javorszky et al., 2015). On-going support involves periodic monitoring of water quality and functionality alongside sustained subsidy from the state to the VWSCs but the support remains largely focused on hardware and technical matters. The Rajasthan SRWSA case study follows the principles of the Swajaldhara policy (Government of India, 2003) while Jharkhand and Chhattisgarh reflect the NRWDP policy (Government of India, 2013a), which has led to some subtle differences with regards to the village-level institutional arrangements for the community service providers which will be explained later in the chapter. However, the enabling support environments follow the same structure.

The Maharashtra case study exhibits traits of an outlier case as it follows a 'hier-archical and technocratic' approach but is one of the richest states. This is partly shaped by the 'developmental' political economy of that state (Kohli, 2012) but it can also be explained by the much more sophisticated technology that is being managed. As opposed to the single-village scheme (SVS borehole) systems in the cases already discussed, in Maharashtra the case study focuses on a multi-village scheme (MVS) for 156 villages and two small towns. Here, the SRWSA take on the role of direct service provider and the role of the community is to establish a VWSC for promoting compliance among the community for regular tariff payment. In this sense, the case study reflects hybrid forms of service delivery which include characteristics of community management and standard utility-type approaches.

The other centralised government support is called the 'Reformed SRWSA' that come from the two Gujarati case studies but these are both focused on the same enabling support environment, called the Water and Sanitation Management Organisation (WASMO). WASMO was born out of a process of sector change in the early 2000s to exist alongside the Gujarati Water Supply and Sanitation Board with a mandate to support the community management of rural water services across the state (Chary Vedala et al., 2015a, 2015b). This led to the development of a state-level organisation that retained a centralised character operating alongside a conventional SRWSA but which integrated both technical support units with 'Social Development Units' that focused on providing specialist information, education and communication (IEC) services to communities to support community management (Chary Vedala et al., 2015a, 2015b). This celebrated programme (Das, 2014; James, 2011) provides an example of how a centralised SRWSA can support a highly participatory model of community management when compared to the other SRWSA cases. The Government of India has attempted to adopt this design principle in the NRDWP through the promotion of Water Supply and Sanitation Organisations at the state level (Government of India, 2013a) but this research found little evidence that such organisations functioned in a meaningful way in any other SRWSA case study.

The three case studies from Sikkim (Saraswathy, 2016c), Tamil Nadu Erode (Saraswathy, 2015) and Kerala Kodur (Chary Vedala et al., 2016b), on the other hand, are all considered to be forms of a decentralised LSG enabling support environment. Before describing these, it is important to clarify that in the centralised SRWSA cases (and all other case studies) the LSG of the Gram Panchayat is prevalent and can play a role either in service provision or support. However, in these three cases the broader Panchayat Raj Institutions are the main support bodies during the service delivery phase and for this reason they are deemed to reflect a decentralised, rather than centralised, form of support system. This is reflected in Figure 10.2 which shows that although a SRWSA still leads the support during implementation through asset creation and other tasks, the primary form of on-going support is channelled through the LSG system. This is considered to reflect the maturity of the 73rd constitutional amendment toward devolution within the rural water supply sector within these states.

Kerala is the standout example of a state that has implemented this devolution agenda (Desai, 2006) and this is reflected in its ranking as the most devolved state as per the Government of India Devolution Index (Government of India, 2015c). The maturity and success of devolution in Kerala is considered to be linked to a number of advantages the state has over others in terms of decentralisation.

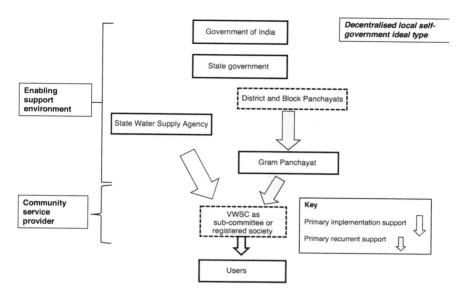

Figure 10.2 Institutional map of decentralised local self-government

Primarily, this includes a well-educated and politically engaged population (Kohli, 2012) but the structure of local administration also means the average Gram Panchayat in Kerala is approximately 50,000 people against a national average of 5,000. This unit of administration means that the 'village-level' LSG units have greater economies of scale and capacity to undertake development works. In the case study from Malappuram district, the Gram Panchayat has a population of 45,000 people and it is this agency that operates as an enabling support environment to a series of beneficiary groups at the habitation level who formed registered societies to become community service providers. Implementation is still undertaken by a SRWSA, in the form of the Kerala Water Authority, but all on-going support is structured through the Gram Panchayat and broader LSG system.

The other two decentralised LSG approaches have an extremely strong role for Gram Panchayats within the service provision tasks and so provide support directly to these institutions from the apparatus of the LSG system. In Tamil Nadu, which ranks sixth on the Devolution Index, this means that although the SRWSA – known as the Tamil Nadu Water and Drainage Board (TWAD Board) – takes on implementation work, the community service provider receives on-going support primarily through the Block Development Officer of the Department of the Panchayat Raj and Rural Development (Hutchings, 2015; Saraswathy, 2015). Similarly, in Sikkim, which comes fourth on the Devolution Index, the Rural Management and Development Department focus on implementation while on-going service delivery and support are provided through the Panchayat Raj structures with the state institution for rural development delivering accredited training to Gram Panchayats to support community management of rural water services with some on-going technical support to water quality management (Saraswathy, 2016c). Together, these decentralised case studies represent government programmes that have moved away from having a centralised SRWSA as the main on-going support agency towards a model where that function has become integrated in local government systems, a set-up that is more common in other low- and lower middle-income countries (Lockwood and Smits, 2011).

Hybrid and external support

Beyond the 'pure' government programmes, there are seven case studies with hybrid enabling support environments and three with external agency enabling support environments. The hybrid case studies include partnerships between SRWSAs and other non-governmental organisations. In the case of West Bengal (Smits and Mekala, 2015) and Madhya Pradesh (Ramamohan Roa and Raviprakash, 2016a) this includes civil society NGOs who can be described as a 'complementary partner' helping to provide services in problematical areas where the standard government model has failed to deliver services. In Tamil Nadu Morappur, the hybrid model also includes a public–private partnership with the private sector delivering and operating a MVS for villages alongside the decentralised LSG model described in the other Tamil Nadu case study (Hutchings, 2015). The final hybrid sub-model is public–donor partnerships with three World Bank supported programmes in Kerala Nenmeni (Saraswathy, 2016a), Punjab (Harris et al., 2016b) and Karnataka (Ramamohan Roa and Raviprakash, 2016b) and one bilateral donor supported pilot programme in Himachal Pradesh (Harris et al., 2016c).

An ideal type institutional set-up for the public–donor partnership is presented in Figure 10.3. It shows that the conventional set-up found within World Bank programmes in India is to form a Project Unit within an existing SRWSA which will receive additional funding from the World Bank and have distinct operating rules to the broader SRWSA, usually including a stronger emphasis on software support and different cost-sharing prescriptions following the principles of the 'demand-responsive approach' to community management (Harris et al., 2016b; James, 2011; Saraswathy, 2016a). This was the case in the Jalanidhi programme, in Kerala, and Jal Nirmal programme, in Karnataka; however an issue with this approach is that there can be a lack of coordination or even conflict between the Project Unit and the broader SRWSA (Ramamohan Roa and Raviprakash, 2016a). The more recently implemented Punjab Rural Water Supply and Sanitation Project has avoided this issue through adopting a sector-wide approach (SWAp) meaning that the SRWSA receives additional funding and capacity building from the World Bank but rolls out the support model across the state, rather than limiting it to a specific project office (Harris et al., 2016b). Despite these differences, the important point in terms of support from the public–donor partnerships (and the hybrid enabling support environments more widely) is that the government

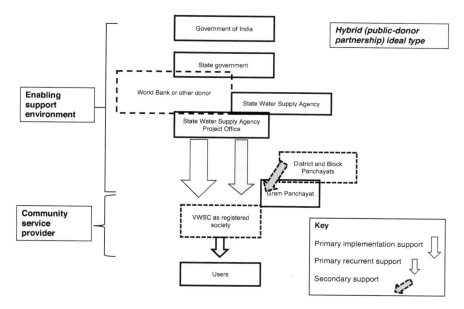

Figure 10.3 Institutional set-up of a hybrid (public–donor partnership) enabling support environment

agencies, usually the SRWSA, remain the dominant partners and are the primary agencies providing support on the ground, but this support is shaped by conditions placed on the SRWSA by donors. The role of the donor is then to provide additional finance as well as capacity building to the SRWSA and then to closely monitor performance.

The final typology of support is classified as external agency support and contains three cases whereby support is provided without the direct involvement of the Indian state (Chary Vedala et al., 2016a; Javorszky et al., 2016; Smits et al., 2016). This is a rare situation in India due to the strong role of the state in developmental activities and the traditional hostility between government and NGOs (Sen, 1999). However, where it exists, there can often be a tiered structure of NGO support as international or even national NGOs often operate through local NGOs, as shown in Figure 10.4. For example, in Uttarakhand this was the case with the Tata Foundation supported Himmotthan Water Supply and Sanitation Initiative which provided support through a local subsidiary known as the Himmotthan Hospital Trust.

There are exceptions to this set up such as the Gram Vikas programme in the state of Odisha in which the NGO works directly at scale with thousands of villages (Javorszky et al., 2016). Either way, the NGO cases tend to provide support in relatively niche situations, such as small, remote Himalayan communities (Smits et al., 2016), fluoride affected areas (Chary Vedala et al., 2016a) or areas dominated

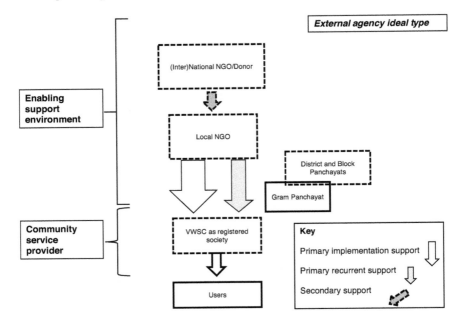

Figure 10.4 Institutional set-up of an external agency enabling support environment

by the tribal castes that have traditionally been poorly provided for by the Indian state (Javorszky et al., 2016). In part due to this selective approach and smaller scale provision, the NGO support can often be much more tailored than the support provided through the government systems. Such differences will now be further highlighted through a comparative analysis of these different enabling support environment typologies.

Characteristics of enabling support environments

The research methodology has developed a number of indicators that can be used as a means for comparing and categorising the characteristics of these typologies. The critical indicator in terms of validating the level of success in rural water services is the service-level outcomes and so this section starts by comparing the outcomes across the categories, with Figure 10.5 showing the results. A proportionally higher number of people in the government programmes achieve basic or above service levels on the composite service-level indicator. The decentralised LSG cases achieve the highest score with 67 per cent of people receiving basic or above service levels while in the centralised SRWSA the proportion of people reaches 63 per cent. The hybrid and external agency categories, on the other hand, have slightly lower outcomes at 56 per cent and 54 per cent respectively. This is partly explained by the inclusion within these groups of case studies from the most

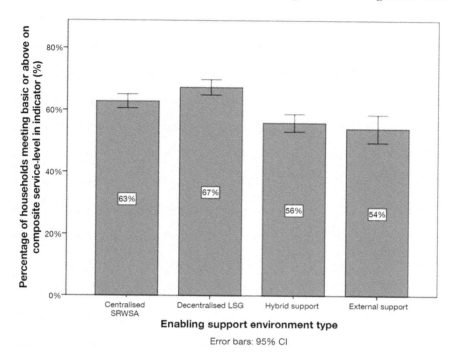

Figure 10.5 Percentage of survey respondents achieving basic or above on the composite service-level indicator by enabling support environment type (programme villages only: $n = 1,732$)

challenging contexts, such as the delta regions of West Bengal that have salinity in groundwater (Smits and Mekala, 2015) or the quality-affected areas of Madhya Pradesh (Ramamohan Roa and Raviprakash, 2016a), which drag the average down. Overall, though, there are not large differences between the enabling support environment types in terms of service-level outcomes.

A key concept associated with the conventional community management approach (and rural water services more broadly) is participation (Harvey and Reed, 2006; Jones, 2011; Kleemeier, 2000; Marks and Davis, 2012). Measured in this research as the level of decision-making power communities have through the service delivery cycle, Table 10.2 shows the median ranking for the 'implementation' and 'service delivery' stages across the categories. It shows some subtle differences across the cases with lower levels of participation in the centralised SRWSA cases compared to the decentralised LSG cases. However, the most significant difference is considered to be in the ranking at the service delivery phase. In the government programmes this is ranked as either functional or interactive while the hybrid and external typologies have a ranking of self-mobilisation. This is thought to reflect the Government of India moving away from the ideal that

Table 10.2 Participation categorised ranking across the enabling support environment (mode)

Category	Participation in implementation	Participation in service delivery
Centralised SRWSA	3 (Functional participation) – 'The community is provided with a detailed implementation plan that they discuss and they have a chance to amend limited elements.'	3 (Functional participation) – 'The community is provided with administration, management and operation and maintenance arrangements that they discuss and they have a chance to amend limited elements.'
Decentralised LSG	5 (Self mobilisation) – 'The community practices self-supply and seeks to improve this, or have developed an implementation plan and seek external support.'	4 (Interactive participation) – 'The community in partnership with the service provider and/or support entities engage in joint decision-making regarding appropriate arrangements for administration, management and operation and maintenance.'
Hybrid	4 (Interactive participation) – 'The community in partnership with the service provider and/or support entities engage in a joint analysis of implementation options before developing a plan.'	5 (Self mobilisation) – 'The community take responsibility for administration, management and operation and maintenance, either directly or by outsourcing these functions to external entities.'
External agency	4 (Interactive participation) – 'The community in partnership with the service provider and/or support entities engage in a joint analysis of implementation options before developing a plan.'	5 (Self mobilisation) – 'The community take responsibility for administration, management and operation and maintenance, either directly or by outsourcing these functions to external entities.'

communities should be taking responsibility for service delivery and moving towards an on-going partnership approach, in the case of the decentralised LSG, or one in which government retains the majority of control, as is the case with the centralised SRWSAs. In comparison, the hybrid and external agency approaches tend to retain an 'international character' that reflects the conventional view that groups of private citizens in the form of a VWSC should take responsibility for service delivery tasks. These findings are considered to indicate a broader point about how community management has evolved in India to become increasingly embedded within the local government institutional framework.

An important argument to improve outcomes within the community manage-ment plus paradigm is the need to drive the professionalisation of rural water services and move beyond an approach reliant on volunteerism and poorly trained individuals (Lockwood and Smits, 2011; Moriarty et al., 2013). As explained in the methods, the consolidated professionalisation indicator used to assess this was based on the mean score of Qualitative Information System (QIS) data-processing tools. In Table 10.3 the mean indicator data has been presented in raw form and re-categorised into the five-point-scale that the QIS follows using basic rounding. It shows the level of professionalisation is categorised as high across all the enabling support environment categories, although the mean indicators that have the highest level are found in the hybrid and decentralised LSG cases.

The research also tried to characterise the relationship between the enabling support entities and the community service provider. For this purpose the partnering assessment tool was used, with Table 10.4 displaying the partnering type that was most strongly highlighted for each category of cases. For the external agency category the partnering type was highlighted as operational, reflecting that communities contribute labour and resources together with the enabling support entities. For example, in the case study from Odisha (Javorszky et al., 2016), it is common for the VWSCs to organise community members to contribute labour to the construction of new schemes, which are managed by the NGO Gram Vikas. The difference in partnering types across the other categories reflects the transac-

Table 10.3 Professionalisation indicator for enabling support environment and community service provider

| Category | Professionalisation of enabling support environment | |
	Consolidated indicator (out of 100)	*Categorisation*
Centralised SRWSA	63	High professionalisation
Decentralised LSG	75	High professionalisation
Hybrid	77	High professionalisation
External agency	65	High professionalisation

Table 10.4 Partnering assessment between enabling support environment and community service provider

Category	*Dominant partnering mode in service delivery*
Centralised SRWSA	Transactional – The ESE and CSP fulfil different elements of the administration, management, and operation and maintenance functions as per negotiated arrangements
Decentralised LSG	Collaborative – The ESE and CSP share responsibility for decisions regarding administration, management, and operation and maintenance
Hybrid	Transactional – The ESE and CSP fulfil different elements of the administration, management, and operation and maintenance functions as per negotiated arrangements
External agency	Operational – The ESE and CSP work together contributing labour and/or resources to support administration, management, operation and maintenance

tional approaches taken to structure relationships in the centralised SRWSA and hybrid approaches. This reflects 'low' levels of partnering in which there is a distinct division between organisational responsibilities and not much joint working, but is considered appropriate for scaled programmes covering many hundreds of villages. The decentralised LSG cases are ranked as having a collaborative partnering typology in which the enabling support environment and community service provider share responsibility and take joint decisions regarding service delivery. This is considered to reflect the greater capacity that the community service providers have in the decentralised case studies.

The final research processing tool which was used to assess organisational elements of the case studies was the adapted Institutional Assessment Tool (Cullivan et al., 1988). For the purpose of the synthesis, Table 10.5 presents the organisational area which was ranked as the strongest and weakest for each category of enabling support environment. It shows that the centralised SRWSA are ranked as having strong technical capability but weak organisational culture. This is considered to reflect the 'hierarchical and technocratic bureaucracy' described by Hueso and Bell (2013, p. 1013) that is designed to deliver physical infrastructure but has an organisational culture that is too static and rigid to adapt to providing broader support functions as the sector has changed. The decentralised LSG category comes out as scoring well on leadership, which is a characteristic associated with successful organisational performance across numerous sectors (Lieberson and O'Connor, 1972). The lowest ranked trait is autonomy, which again is considered to be reflective of organisations embedded within the inflexible bureaucracy of the Indian state. While potentially framed as a criticism, it is noted that bureaucratic process can be a very powerful force for promoting standardisation (and therefore consistency) for enabling support entities that can have responsibility for many tens of millions of people, which is often the case in India. Finally, the external agencies have strong organisational culture reflecting the socially oriented character of many NGOs, but are limited in terms of technical capability, often as they focus on software elements rather than hardware.

The various organisational-based assessment tools help distinguish between the characteristics of different enabling support environment types. Recognising the possible fallibility of specific tools, it is when triangulating these types of results together that it is possible to identify patterns across the different types of support. Focusing on the government support approaches, many of the measures support a more positive picture of the decentralised LSG model compared to the centralised SRWSA. It has a 'collaborative' partnership approach between entities that both rank in the 'high professionalisation' category, with higher levels of community participation and ultimately better overall service-level outcomes. The research cannot say conclusively whether one of these aspects has a causal link to the other but, it is argued, these types of processes are likely to be synergistic, creating positive feedback loops between different elements to deliver better overall outcomes. It should be noted that the decentralised LSG cases tend to be found in richer states with better governance capacity and in this sense are likely to reflect the broad capacity to deliver services effectively that appears to comes with greater

Table 10.5 Institutional assessment outcomes

Category	Strongest organisational area	Weakest organisational area
Centralised SRWSA	Technical capability – is the measure of the institution's competence in conducting the technical work required to carry out the responsibilities of the institution.	Organisational culture – is the set of values and norms which inform and guide everyday actions. The culture forms a pattern of shared beliefs and assumptions which translate into behaviour which can be observed.
Decentralised LSG	Leadership – is the ability to inspire others to understand the institution's mission, to commit themselves to that mission, and to work toward its fulfilment.	Organisational autonomy – is the institution's degree of independence from the national government or other governmental or regulatory bodies. While not unrestrained, this independence must exist to the extent that the institution is able to conduct its affairs and meet its responsibilities in an effective manner with minimum interference and controls by other entities.
Hybrid partnership	Technical capability – is the measure of the institution's competence in conducting the technical work required to carry out the responsibilities of the institution.	Leadership – is the ability to inspire others to understand the institution's mission, to commit themselves to that mission, and to work toward its fulfilment.
External agency	Organisational culture – is the set of values and norms which inform and guide everyday actions. The culture forms a pattern of shared beliefs and assumptions which translate into behaviour which can be observed.	Technical capability – is the measure of the institution's competence in conducting the technical work required to carry out the responsibilities of the institution.

wealth and development as was explored earlier in this book. The WASMO case studies from the wealthy state of Gujarat also support this hypothesis as, although they follow a centralised approach, they have adopted what has been described as a reformed SRWSA approach and have better outcomes across the various organ-isational-based assessment tools as well as service levels, compared to the poor states following a centralised model.

Further developing the institutional analysis, this section ends by considering the financial differences between the enabling support environment typologies. As will be examined in greater detail during the following chapter, there is a high degree of variability between case studies in terms of financial costs. This is reflected in Table 10.6 that shows the interquartile range for total CapEx and recurrent costs (OpEx, OpEx Enabling Support and CapManEx) by the enabling support environment types. With the ranges overlapping for CapEx across all categories there are no clear conclusions to draw from the table in terms of the level of total implementation-related investment across the typologies. This suggests that the enabling support environment categories do not tend towards different levels of costs and, as examined in the next chapter, that other factors are more important in terms of differentiating CapEx.

For recurrent costs, however, the interquartile range for the local self-government supported case studies shows higher levels of recurrent investment than the other cases. The lower quartile for OpEx in that category is higher than the upper quartile reported for SRWSA and external agency cases, indicating different investment patterns between these enabling support environment types. The hybrid-supported cases fall in-between with higher levels of recurrent investment than the SRWSA and external agency but lower levels than the local self-government. In Chapter 2, it was argued that recurrent financing was an area of underinvestment in the rural water sector (Burr and Fonseca, 2013; Franceys et al., 2016). The data presented in Table 10.6 indicate that the local self-government case studies receive the highest level of recurrent expenditure on average. In the following chapter further evidence is presented on how these costs, and others, are split between the support entities themselves and the community, helping to provide clarity of the issue of cost sharing found in successful cases from India.

In summary, this section has highlighted a number of trends in the organisational arrangements for community management in India. It has shown a difference in the pattern of emphasis between programmes which are considered to be domestically-driven (i.e. Government of India) and those which are more internationally driven (i.e. hybrid and external agency). It has been argued that this is leading to a split between programmes that retain a conventional community management approach and a shift to programmes that have greater sharing of responsibility between support agencies and communities. The following section will develop these ideas by focusing on a review of the community service provider arrangements found across the case studies.

Table 10.6 Enabling support environment typology overall costs

Enabling support environment type	CapEx		Recurrent	
	LQ	*UQ*	*LQ*	*UQ*
SRWSA	$85	$208	$6	$11
Local self-government	$128	$251	$28	$73
Hybrid	$166	$342	$10	$46
External agency	$38	$279	$2	$16

Community service providers

This section advances the analysis of organisational arrangements to focus on the community service provider level. Again, the intention here is to provide a synthesis on the *types* and *characteristics* of the community service providers across the case studies. There are considered to be two main types of service provider found across the case studies that can be classified as 'community management through local self-government' and 'community management through societies'. Across the classifications the entities that take on the service delivery tasks are most commonly referred to as village water and sanitation committees (VWSCs) and, although there are some variations, VWSC will be used to describe such a body in all circumstances. As will be explained there is a division within each classification regarding the status of the VWSCs. There is a split between what are described as 'VWSCs as autonomous sub-committees of the local self-government' and 'VWSCs as representative sub-committees of the local self-government' and then a distinction between 'VWSCs as registered societies' and 'VWSCs as unregistered societies'. Table 10.7 shows the allocation of cases and then each typology is described below.

Community management through societies

The community management through societies approach is presented first, as this approach is most reflective of conventional ideas outlined as per the international understanding of community management. There are two cases that will be used to illustrate this which come from different ends of the technical spectrum. Case study Kerala I from Malappuram district is an example of a registered society managing a SVS piped water supply scheme from a surface water source. The other case study discussed is from West Bengal and involves an unregistered society managing handpumps. First it is useful to explain the importance of the Indian

Table 10.7 Overview of case studies by type of enabling support environment

VWSCs as registered societies	VWSCs as unregistered societies	VWSCs as autonomous sub-committees of the local self-government	VWSCs as representative sub-committees of the local self-government
3. Odisha	2. Madhya Pradesh	6. Rajasthan	1. Jharkhand
8. Telangana	7. West Bengal		4. Chhattisgarh
12. Uttarakhand	10. Himachal Pradesh		5. Meghalaya
11. Punjab			9. Karnataka
13. Kerala I			17. Tamil Nadu I
14. Kerala II			18. Tamil Nadu II
15. Gujarat – Gandhinagar			19. Maharashtra
16. Gujarat – Kutch			20. Sikkim

Societies Registration Act, which is a piece of legislation from the colonial era, administered by the Ministry of Corporate Affairs (Government of India, 1860). All charitable bodies (as well as scientific and literary societies) should be registered under the Act, which means notifying the state of at least seven board members and the agreed upon rules and regulations of the society. A registered society can then open a bank account and have official (i.e. contractual) agreements with government entities. They retain, however, independence from government as they are a distinct organisational body. VWSCs that are not the sub-committee of the Gram Panchayat should be registered under this model.

The described principles can be seen in the Kerala I case study (and six other cases outlined in Table 10.7). In the Kerala example, VWSCs are formed at the habitation level by community members to take on the management of rural water services. The VWSCs are registered societies so have an independent bank account and an established set of rules and regulations for managing the service. The relationship with the local self-government is arranged through a Memorandum of Understanding (MoU) that sets out the division of responsibility between the VWSCs and the local self-government. In Kerala, the registered society is supported by an enabling support environment that was classified as a form of decentralised local self-government support in the section above. This situation emerged partly due to the larger administrative boundaries found in Kerala, which makes the Gram Panchayats at a scale whereby they can provide support to a number of different VWSCs, rather than being at the scale of a single VWSC.

The case studies also contain examples of unregistered societies and in such situations the broad institutional model of relationships is similar to the registered society approach. The case study from West Bengal, for example, is of VWSCs that are established to manage single handpump installations. The professionalisation of these VWSCs is comparatively low compared to other cases and they follow a reactive approach to repairs by collecting money as need arises and hence do not require a bank account and registered status (Smits and Mekala, 2015). Again, the Gram Panchayat is part of the enabling support environment rather than the service provider and it receives support from a complex of NGOs and private pump mechanics to aid it with these tasks. This type of unregistered society is rare in India and is only found in this research in three cases where NGOs or donors have operated (Himachal Pradesh, Madhya Pradesh and West Bengal).

Either way, the institutional architecture of community management through societies is outlined in Figure 10.6. This shows that the set-up follows the classical ideas for public service delivery set-up by the accountability triangle (World Bank, 2004). This means there is independence between the service provider and the oversight functions within the system. As illustrated in the diagram this means there are two routes for users to hold the service provider to account, which are described as 'route of local service responsiveness' in which community members can go directly to the registered society if there is a problem and the 'route of local service oversight' in which the users can go to the local self-government who have the power to effectively revoke the right of the registered society to continue their role of service provider. In this way and under this set-up, the village-level local

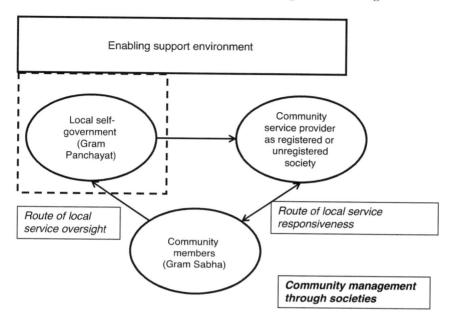

Figure 10.6 Formal routes of accountability in the 'community management through
societies' model

self-government is part of the enabling support environment rather than service
provider.

Community management through local self-government

Although some form of registered or unregistered society was the service provider
in 12 of the cases, the Constitution of India mandates that rural water services are
the responsibility of the local self-government of the Gram Panchayat
(Constitution of India, 1950). Under the Swajaldhara (Government of India, 2003)
and later NRDWP (Government of India, 2013a) policy programmes, the turn to
community management sought to reconcile this promotion of private citizens
taking on the role of service provision with this constitutional requirement for
local self-government playing a role, through requiring VWSCs to be established as
sub-committees of the local self-government. This means a common model for
community management within government programmes is one in which the
VWSC is embedded within the broader institutional architecture of the local
self-government system of the Panchayat Raj. However, this research has indicated
that there has been a shift in the more recent NRDWP (Government of India,
2013a) between the extent to which VWSCs operate as an autonomous body of

the broader local self-government, to a situation in which the local self-government becomes the direct service provider and the VWSC plays a representative role. This has led to the development of two sub-categories: the 'autonomous VWSC' and 'representative VWSC'.

Having an autonomous VWSC, which is not a registered society as described in the previous section, is limited to just one case but is labelled as a sub-category to help illustrate the change in the government approach between its two most recent policy programmes. The Rajasthan case study was part of the Swajaldhara (Government of India, 2003) programme under which the VWSC remains a sub-committee of the Gram Panchayat but operates largely in an independent manner, having a clear division of responsibility between the VWSC and broader Gram Panchayat. This partly reflects the greater emphasis the Swajaldhara placed on establishing and training VWSCs as a condition for receiving government support. However, as the report on the Rajasthan case study makes clear (Harris et al., 2016a), the VWSCs in this case can be more accurately described as isolated rather than autonomous. The policy design is considered to reflect the greater international influence on the ideas of community management that were incorporated into the first nation-wide community management programme, compared to the later NRDWP.

What is described as the representative VWSC model is now more common across the government case studies with this model found in Chhattisgarh, Jharkhand, Meghalaya, Maharashtra, Tamil Nadu and Sikkim. In these examples,

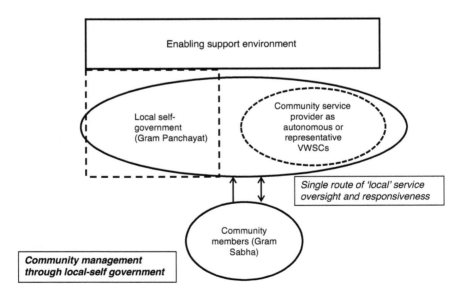

Figure 10.7 Formal routes of accountability in the 'community management through local self-government' model

VWSCs are formed but the effective operation of the committees is completely dependent on the Gram Panchayat of the local self-government. With the VWSCs mandated to make key members of the committees the most powerful figures from the local self-government it means the everyday operation of the systems are controlled by the Gram Panchayat office. This is considered to bring both benefits and dangers. It means that the institutional resilience of the system is much stronger as the local self-government is a permanent organisation that is financially supported through the broader taxation system. However, it does provide dangers regarding the conventional ideas about the local lines of accountability for this public service.

As indicated when comparing Figures 10.6 and 10.7, having the VWSC as the sub-committee of local government means that that the service provision and oversight tasks become integrated into the same institution. At a single village level there is not the organisational capacity to have an executive division of the government providing services and a legislative/regulatory branch that holds the executive to account. This means it could be argued this set-up is less accountable to the users as it removes the principle of an independent service provider. This is intentional, however, as according to officials in various interviews and meetings conducted as part of this research, the primary purpose of this institutional design is to simplify the overall governance system and reduce the number of different bodies that operate at the village level. So the idea goes, concentrating power in the Gram Panchayat reduces the potential for conflict between opposing leaders and/or bodies with villages, which has been a barrier to development activities in India due to the 'vibrancy' of local democracy.

Characteristics of the community service provider types

This section now compares the characteristics of the community service provider arrangements starting with service-level outcomes. What it shows is that, contrary to arguments about the importance of the enabling support environment, it is the type of community service provider arrangements that shows the biggest differences between sub-groups. There is a group of cases that are clearly more successful in terms of service levels, which are those where the service provider is a registered society, where 80 per cent of the population receive services meeting all the government norms. The next best performing category is the representative VWSC model, which is considered the 'standard' Government of India approach to community management following the NRDWP. Here, over half the people receive services meeting all government norms and the median service level is basic. The unregistered society and autonomous VWSC categories are much rarer across the study and have the worst outcomes from any type of service provider arrangements.

The indicator for the professionalisation of service provision tasks follows a similar pattern to the service level outcomes, as shown in Table 10.8. This indicates that professionalisation of service provision could be an important factor in whether services are successful. The requirements around registration under the Societies Act and also the requirements for representative VWSCs as sub-committees of the local

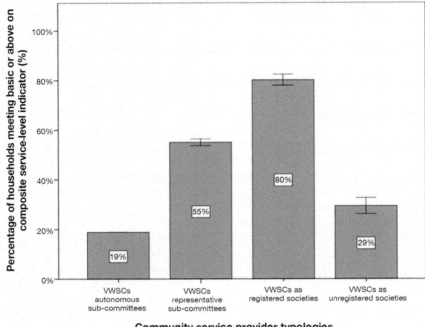

Community service provider typologies

Error bars: 95% CI

Figure 10.8 Percentage of survey respondents achieving basic or above on the composite service-level indicator by community service provider type (programme villages only: *n* = 1,732)

self-government, are considered to at least be aligned, if not helping to drive, the higher professionalisation in these groups. The unregistered societies are by definition going to score lower on the professionalisation indicator that reflects partly the formal mandate that the entity has to deliver services. It is hard to draw too many conclusions about the autonomous VWSC model as it is only reflecting one case study from the older Swajaldhara programme. The differences in participation are more subtle and harder to interpret, but they do not appear to be as critical to service-level outcomes as professionalisation at this level.

As was completed above for the enabling support environment categories, the broad levels of financing reported across the community service provider typologies is presented here. Table 10.9 shows that there is again significant variability in the CapEx costs in the categories apart from the single case study category of the autonomous VWSC. The registered society case studies have markedly higher CapEx, with a lower quartile starting at $176 per person. With this category also enjoying the standout results in terms of service levels, the data could indicate a

Table 10.8 Participation and professionalisation of community service provision types

Type of community service provider	Participation		Professionalisation	
	CapEx	OpEx	Score	Categorisation
Registered society	5 (Self-mobilisation) – 'The community practises self-supply and seeks to improve this, or have developed an implementation plan and seek external support.'	4 (Interactive participation) – 'The community in partnership with the service provider and/or support entities engage in joint decision-making regarding appropriate arrangements for administration, management and operation and maintenance.'	73	High professionalisation
Unregistered society	4 (Interactive participation) – 'The community in partnership with the service provider and/or support entities engage in a joint-analysis of implementation options before developing a plan.'	4 (Interactive participation) – 'The community in partnership with the service provider and/or support entities engage in joint decision-making regarding appropriate arrangements for administration, management and operation and maintenance.'	48	Medium professionalisation
Autonomous VWSC	3 (Functional) – 'The community is provided with a detailed implementation plan that they discuss and they have a chance to amend limited elements.'	3 (Functional) – 'The community is provided with administration, management and operation and maintenance arrangements that they discuss and they have a chance to amend limited elements.'	33	Low professionalisation
Representative VWSC	4 (Interactive participation) – 'The community in partnership with the service provider and/or support entities engage in a joint-analysis of implementation options before developing a plan.'	4 (Interactive participation) – 'The community in partnership with the service provider and/or support entities engage in joint decision-making regarding appropriate arrangements for administration, management and operation and maintenance.'	61	Medium professionalisation

Table 10.9 Community service provider types overall costs

Community service provider types	CapEx		Recurrent	
	LQ	*UQ*	*LQ*	*UQ*
Registered society	$176	$263	$8	$31
Unregistered society	$38	$342	$2	$10
Representative VWSC	$99	$266	$6	$37
Autonomous VWSC	$93	$93	$11	$11

relationship between higher overall CapEx costs and service-level outcomes. Such relationships are considered in Chapter 11, which examines the correlations between service-level outcomes and financing levels. For recurrent costs there is again some overlap between the interquartile ranges for all the community service provider categories, however the upper quartile for the unregistered society and the single data point for the autonomous VWSCs are markedly lower than the other two categories. This indicates that higher ranges of recurrent investment were found in the registered society and representative VWSC models, which also had the most successful outcomes in terms of service levels.

This section now ends by briefly showing how these classifications match across the analytical levels, with an overview given in Table 10.10. It shows that within this study there is great diversity of potential organisational set-ups and that even within a specific category of enabling support environment there are different community service provider set-ups depending on the case study. This shows that there is adaptability within enabling support environments in terms of the type of

Table 10.10 Matching of organisational arrangements across the case studies

Type of community service provider	Autonomous VWSC	Representative VWSC	Registered society	Unregistered society
Centralised SRWSA	6. Rajasthan	1. Jharkhand, 4. Chhattisgarh 5. Meghalaya 19. Maharashtra	15. Gujarat I 16. Gujarat II	–
Decentralised LSG	–	17. Tamil Nadu I 20. Sikkim II	14. Kerala II	–
Hybrid	–	18. Tamil Nadu II 9. Karnataka	11. Punjab 13. Kerala	2. Madhya Pradesh 7. West Bengal 10. Himachal Pradesh
External agency	–	–	3. Odisha 7. Telangana 12. Uttarakhand	–

service providers they can support. Saying this, there are also some broad patterns of institutional matching which can be highlighted. The concept of an autonomous VWSC sub-committee of the local self-government is a limited model for service provision which can be described as a legacy of a previous policy programme so it not relevant in this discussion. The representative VWSC approach, however, is the most common model for government-run enabling support environments types. It can be described as the standard model in India that reflects the nexus of the local self-government and VWSCs at the community service provider level. However, this analysis shows it can be supported by either a centralised agency in the form of a SRWSA, within the decentralised local self-government model or the hybrid approach.

The external agency-led enabling support environments operate beyond the local self-government system and therefore representative VWSCs are not found within this approach. Registered societies are even more adaptable and are found across the four support models. As this is the highest-performing service provider form, it is encouraging that it can be adapted to each of the support environments. Within this model, the local self-government can remain involved with rural water services but it moves to becoming part of the (local) enabling support environment rather than as a sub-component of the community service provider. In contrast, the unregistered societies are only found in the hybrid enabling support environment cases where NGOs or donors are part of the support environment. In Madhya Pradesh and West Bengal NGOs have worked to establish management committees at sub-village level with this smaller scale considered to lend itself to the unregistered VWSC models. Similarly the scale of management in Himachal Pradesh is also at the sub-administrative village level and hence is considered to play a role in this informality. Overall this section has compared the performance of the different community service provider arrangements and has considered how they match the enabling support environment types, showing that there is much diversity in the patterns of institutional matching.

Conclusions

This chapter has shown that the research has covered a variety of different organisational arrangements for community management in India. This is considered to reflect the effective decentralisation of rural water services from the federal government to the state level, while the diversity between states can be partly explained by the varied level of *intra*-state decentralisation. Two major typologies of government enabling support environments were identified: the centralised SRWSA and the decentralised LSG models. The centralised model, the most common approach, was linked to the on-going legacy of supply-driven rural water services while the decentralised local self-government approach was reflective of the effective maturity of the devolution processes in Kerala, Tamil Nadu and Sikkim, where the on-going support for service delivery has been transferred from the SRWSA to the local self-government institutions. Even among the hybrid approaches the strong role of the SRWSA and local self-governments was stressed.

Recognising that the cases were purposively selected, and are not therefore representative of all rural water services across India, the prevalence of government support across the case studies is still considered to show the strong role of the state in rural water services in India, and thus the comparatively weaker role of civil society and the private sector. This is of relevance when considering the historical development of the community management model around the world which emerged in part due to NGOs and donors attempting to bypass failing (local) governments (Harvey and Reed, 2006; Moriarty et al., 2013). The situation documented in India is completely distinct from this idea of community management as a way to bypass government. Rather the case studies here show that government – in its different forms – is the primary agent supporting community management. Building on this theme, at the community service provider level, this chapter has highlighted the overlap and tension between the concept of community management as a form of autonomous – effectively private – management of water services by citizens and the reality of its practice in modern India which includes significant sharing of responsibility between state and citizen. This theme of 'coproduced rural water supply' is explored more fully in the final chapter.

11 The cost of good services

This chapter progresses the research to consider the service level and financial cost data. It is designed to verify the level of success found across the case studies, as measured through the household surveying, and to interrogate how much it costs to deliver the types and levels of service found. The chapter provides a contribution to the literature on the financial sustainability of rural water services (Burr and Fonseca, 2013; Hutton and Varughese, 2016; McIntyre et al., 2014) by providing guidance on the levels of investment found in successful community management programmes in India.

Structurally, it first focuses on the service levels across the cases to provide a descriptive analysis of outcomes at the whole sample level before focusing on individual cases and groups of cases. This is followed by the financial analysis that follows a similar pattern of moving from the whole sample to case-by-case data before analysis by different sub-groupings of cases. The chapter ends by bringing these two elements together to consider whether it is possible to identify patterns of financial costing arrangements that are associated with certain service-level outcomes.

As with the previous chapter, an overview of key results is presented overleaf in Table 11.1 to provide a reference point for readers to use alongside the case study summaries via the project website.

Service levels across the sample

The analysis of service levels begins by presenting an overview of the composite service-level indicator for both programme and control villages across the entire sample. Although the data distribution shares a similar U-shape, as shown in Figures 11.1 and 11.2, there is a clear difference in the skewness of the data, with proportionally over twice as many households in the programme villages receiving high service levels as compared to the control villages. Based on the research methodology's definition of success, as measured by the composite service-level indicator, and the use of median as the appropriate central tendency measure for ordinal data, this difference in skewness is reflected in the median service level in programme villages being 'improved' (4) compared to 'sub-standard' (2) in the control villages. The Mann Whitney Test confirms the difference in medians as

Table 11.1 Summary of key service level and financial findings by case study

Case	State	Service level (median)	Percentage of population reaching basic or above service level	Capital expenditure (CapEx)	Percentage support contribution to CapEx	Annual recurrent costs	Percentage support contribution to recurrent costs
1	Jharkhand	Sub-standard	31%	$208	100%	$6	71%
2	Madhya Pradesh	No service	35%	$166	100%	$10	3%
3	Odisha	High	69%	$169	82%	$4	10%
4	Chhattisgarh	Basic	55%	$112	100%	$3	26%
5	Meghalaya	Sub-standard	45%	$85	95%	$6	61%
6	Rajasthan	Sub-standard	19%	$93	89%	$11	15%
7	West Bengal	No service	0%	$38	98%	$2	77%
8	Telangana	Improved	87%	$279	88%	$16	59%
9	Karnataka	Sub-standard	37%	$282	95%	$12	4%
10	Himachal Pradesh	High	65%	$342	97%	$6	3%
11	Punjab	High	98%	$247	94%	$50	46%
12	Uttarakhand	No service	0%	$536	91%	$13	54%
13	Kerala I – World Bank	High	100%	$184	83%	$30	52%
14	Kerala II – Local self-government	High	94%	$221	97%	$32	30%
15	Gujarat – WASMO Gandhinagar	Improved	87%	$73	92%	$6	73%
16	Gujarat – WASMO Kutch	High	98%	$196	91%	$9	52%
17	Tamil Nadu – Local self-government	Improved	63%	$128	90%	$28	75%
18	Tamil Nadu – Public–private hybrid	Improved	53%	$17	91%	$46	72%
19	Maharashtra	Improved	94%	$1,019	100%	$13	53%
20	Sikkim	Sub-standard	45%	$251	98%	$73	94%

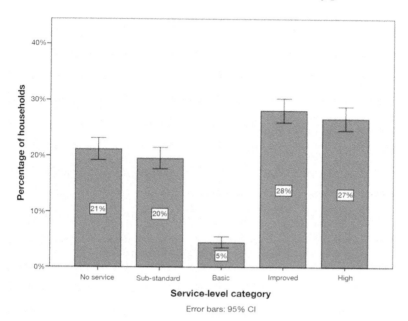

Figure 11.1 Consolidated service-level indicator (programme villages)

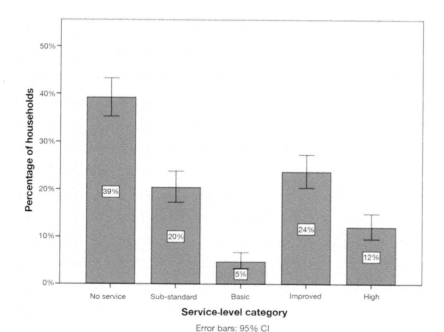

Figure 11.2 Consolidated service-level indicator (control villages)

statistically significant at the 0.00 level. This verifies that at the whole sample level the study compared more successful service delivery in the programme villages than in the control villages. However, even for programme villages, over 40 per cent of the sample report having water services that score 'sub-standard' or 'no service' with regards to the composite indicator. This indicates that within these purposively selected high-performance programmes there are still a significant minority of the population receiving inadequate services. In the WASHCost research it was reported that the 'majority' of people with improved-water sources do not receive a 'basic' service level (Burr and Fonseca, 2013). Here, the results are at least reversed in that the majority of households (54 per cent) in programme villages *do* achieve 'basic' or above service levels. However, the research can be considered to confirm a pattern of results found in other studies – that the improved-water source access figures can mask variable service levels in rural water service programmes (Clasen, 2012; Godfrey et al., 2011).

Moving onto an analysis of the disaggregate service level it is helpful to understand which of the service-level parameters are having the most influence on the composite indicator. In total, across the whole sample 1,019 (31 per cent) respondents failed on at least one parameter, meaning that due to the nominal logic of the composite indicator they were labelled as sub-standard or no service. Of those failures, 573 failed on just one indicator, 319 failed on two indicators while the rest failed on three or more indicators. There is greatest variability in the parameter for quantity, compared to accessibility, perceived quality, reliability and continuity. The distribution of the data is split for quantity with a median measure of 'high' (5) across the sample but the highest standard deviation of any measure. This is reflected in the number of households failing on that measure with this coming out at 676 (30 per cent of the sample). Accessibility is the next measure that is most likely to be reported as failing to meet the service level (16 per cent) followed by reliability and continuity at 11 per cent. Only 3 per cent of respondents reported unacceptable quality but as explained in the methods this is only a measure of perceived quality so has to be treated with caution. The disaggregated service-level data therefore shows that a focus on increasing the quantity of water supplied will have the biggest impact in terms of moving populations up the service-level ladder from 'no service' or 'sub-standard' to 'basic' or above. Evidence indicates that having a water source in or close to the home increases the quantity of water people consume (Howard and Bartram, 2013) and so, in this sense, the Government of India's drive for household connections is well focused in terms of improving service-level outcomes (Government of India, 2013a).

Shifting the focus to the case-by-case service-level data, the analysis shows that 13 of the cases studies have a median measure of 'basic', 'improved' or 'high' on the composite indicator in programme villages. The services delivered in these cases therefore, on average, meet or exceed the standards for all the individual service-level parameters. From these, there are six case studies that have a median service level of 'high', giving a sub-group of what can be considered very successful service-level outcomes. In the middle there are seven that achieve the 'basic' or 'improved' level, while seven of the case studies have medians that come at either 'sub-standard' or 'no service'.

Based on the research methodology, those latter seven case studies cannot be considered 'successful' rural water services as per government norms. For this research the three groups, as shown in Table 11.2, can be labelled as the 'high service level group', 'basic or improved service level group' and 'sub-standard or no service level group'. This finding can be interpreted in different ways. It shows that community management can play a role in delivering service levels at various ranks of success suggesting it has a role to play in countries progressing from a focus on basic access to increasing service levels. However, it also shows that a third of 'reportedly successful' community management programmes fail to deliver even basic service levels, calling into question the overall approach the sector takes to delivering such services.

Service-level outcomes by sub-groups

This section now considers patterns of service-level outcomes across different groups of cases. It begins by providing some further details linked to the previous chapter by comparing the types of enabling support environment and community service providers. However, it also goes beyond these categories to consider service-level outcomes by technical stratifications and the context of the case study.

Table 11.2 Descriptive statistics on service-level composite indicator (all cases, programme villages only)

Service-level grouping	Case study no.	Case study state	Median (on service level indicator)	Mean (out of five)
High-performance group	14	Kerala II	High	4.97
	13	Kerala I	High	4.76
	11	Punjab	High	4.63
	16	Gujarat II	High	4.57
	3	Odisha	High	3.87
	10	Himachal Pradesh	High	3.78
Medium-performance group	19	Maharashtra	Improved	3.89
	15	Gujarat I	Improved	3.74
	8	Telangana	Improved	3.66
	17	Tamil Nadu I	Improved	3.16
	5	Meghalaya	Improved	3.06
	18	Tamil Nadu II	Improved	2.93
	4	Chhattisgarh	Basic	2.68
Low-performance group	9	Karnataka	Sub-standard	2.69
	20	Sikkim	Sub-standard	2.69
	1	Jharkhand	Sub-standard	2.24
	6	Rajasthan	Sub-standard	2.13
	2	Madhya Pradesh	No service	2.18
	7	West Bengal	No service	1.06
	12	Uttarakhand	No service	1.14

The intention of this analysis is not to identify a single critical stratification but to illustrate how outcomes change across groups so to provide further insight into the conditions that are associated with successful rural water services.

Organisation types

Focusing on the differences between organisational types, as shown in the previous chapter, the differences are more marginal when stratifying the cases by enabling support environment with three categories – SRWSA, LSG and Hybrid – delivering a median outcome of 'improved' (4) while the external support category comes out at the 'sub-standard' (2) level. The lower performance of the external agencies is considered to reflect the role the largely NGO-orientated category plays in the Indian sector – they operate in the most challenging contexts when government supplies have failed and hence the level of service they deliver is likely to be lower. For the community service provider typologies the difference in terms of service-level outcome is more pronounced. The Registered Society as a VWSC has a median outcome of 'high' (5) that shows that this is the organisational type that can most directly be associated with high service-level outcomes. The VWSCs as representative sub-committees of the LSG achieves a 'basic' (3) level of service while the other two categories come out at sub-standard and no service. As these categories were the subject of extensive discussion in the previous chapter they are not discussed in detail here, but it was considered appropriate to present the descriptive statistics within this chapter and within the broader statistical analysis.

Types of water-source access and system design

The pattern of service-level outcomes by system design and the type of water-source access are now considered. Figure 11.3 helps illustrate how services levels vary by system design with what is considered to be a distinction between three overall groups. Non-piped water supply, in this study from a single case containing a handpump, delivers universally unacceptable service levels. The SVS from open wells and gravity-fed systems delivered services where the majority of survey respondents reported service levels that did not meet the basic criteria. This is considered to be because of the use of public stand-posts in such contexts (Ramamohan Roa and Raviprakash, 2016a; Saraswathy, 2016b, 2016c). The other types of piped supply delivered services where the majority of the population received acceptable service levels.

As well as the technical system design, the household survey recorded the type of water-source access available to respondents. This could be a household tap connection, public stand-post or handpump, for example. The sub-sample sizes for these categories indicate that the sample predominately covered services provided through household connections rather than communal sources. With Indian policy focusing on piped water supply (Government of India, 2013a) the data shows that household connections deliver significantly better services than public stand-posts (and that even private wells deliver consistently higher services than public

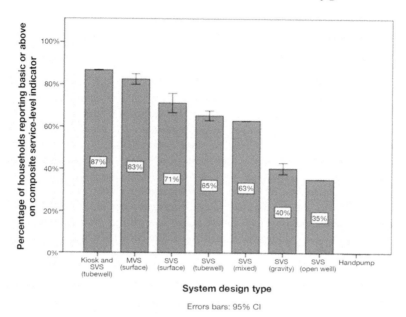

Figure 11.3 Percentage of households meeting at least basic on the composite service-level indicator across water-source types

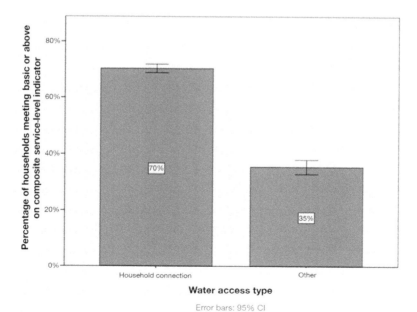

Figure 11.4 Percentage of households meeting at least basic on the composite service-level indicator across water-delivery types

stand-posts). Overall, the data here suggests that the type of water delivery is an extremely important factor in terms of differential performance between the case studies, more so than the technical design of the system.

Broader context of case study

Due to links between broader development indicators and rural water services (as outlined in Chapter 3) it is considered useful to assess the cases by the two broad contextual factors that were related to rural water service outcomes at the state level. It is acknowledged that the case studies cannot be considered representative of state-level trends as they are purposively selected single cases, however the strength of the synergistic relationship between water services and broader societal development mean that this approach is still considered useful. As such, Figure 11.5 shows service-level outcomes are better in the case studies from the richest states (when excluding case studies from the mountain and hilly states which it was previously argued are considered a special case for water services in India). Comparing the middle-income to high-income aggregate service-level data also

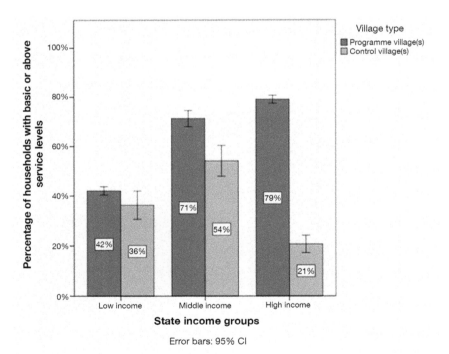

Figure 11.5 Percentage of households meeting at least basic on the composite service-level indicator across wealth categories

reinforces the arguments made about the social-democratic tendencies versus development tendencies which tend to be found in the middle-income and high-income states (Kohli, 2012), respectively. Control villages in the middle-income states have better service-level outcomes than programme villages reported on in low-income and mountainous states, whereas the control villages in the development states have the lowest overall outcomes from across the whole study. This would support the connections the book makes between the political economy of the state and rural water service outcomes. The Maharashtra case study, for example, is considered to reflect an example whereby the state government has taken a top-down approach to constructing large MVS piped supply, which delivers strong service-level outcomes to the villages it serves. However, beyond these large schemes, the 'forgotten' villages seem to lack the capacity or enabling environment to deliver good service themselves, whereas in Kerala, for example, even for villages outside the main programmes studied, widespread literacy and development appears to allow villages to deliver services adequately.

The difference between the service-level outcomes in the case studies from the mountainous and hilly states versus the rest is less clear. In the programme villages, the rest of the sample outperforms the mountainous case studies but the control villages have better outcomes than the programme villages in the mountainous villages (although the difference is within the margin of error). This may reflect a sampling bias for the control villages in the mountainous states. In conversation with the field-researchers for these case studies the remoteness of the programme villages was mentioned, specifically in relation to the Uttarakhand and Himachal Pradesh case studies (Harris et al., 2016c; Smits et al., 2016). Selecting control villages that were as remote as the programme villages may have been due to a geographically oriented sampling bias with researchers selecting control villages that were closer to the main roads. Similar sampling bias has been reported in other development research projects (Chambers, 2008).

Service-level overview

The analysis of the service-level data from the household surveying has revealed that there are statistically significant differences across various sub-groups of the sample that support the theoretical justifications for selecting the case studies. This includes higher service levels in programme rather than control villages, differences between economic income categories, technical mode of supply and organisational types. It is not argued, nor was the analysis intended to test, whether any of these factors are causal variables as the research understands rural water service outcomes to be shaped by complex pathways of causality. Rather the descriptive analysis was designed to validate the level of success across the cases and provide empirically grounded insights into the conditions associated with success that can be used to further structure, and triangulate with, our broader understanding of community management in India.

Financial analysis

This section begins the process of moving the quantitative analysis from the validation measures of the study (i.e. service levels) to one of the critical elements for answering the research questions (i.e. financial flows). Similarly, it begins by providing an overview of the general financing pattern across the major cost categories. It then moves to a cross-case analysis that disaggregates the cost-sharing arrangements between communities and external support entities for each case study. These patterns are then analysed by the major groupings used in the previous section including organisational types, technical modes of supply and economic income categories. The financial data was consolidated to the case level based on the average of three programme villages at the community service provider level and the costs collected at the enabling support environment level. Inconsistent data was collected at the enabling support environment level for control villages so the financial analysis does not compare programme against control data. With this section focusing on financial data, the basic central tendency measure used is the mean and the interquartile range (IQR) as opposed to the median in the section before. With significant variability in interval data this is deemed an appropriate approach (Black, 2012) – especially as this replicates the approach followed in the WASHCost research (Burr and Fonseca, 2013; Fonseca et al., 2011; McIntyre et al., 2014) enabling direct comparison between the data as the analysis is presented.

Financial costs across the sample

There is considerable variability across the cases in terms of the level of financial investment found. As shown in Table 11.3, at the whole sample level the average CapEx on hardware was $207 per person with an IQR of $93 to $233, while the mean CapEx on software was $25 with an IQR of $0.7 to $17. The mean recurrent costs come to around $20 per person but again there was variability across the programmes. For OpEx and OpEx Enabling Support this research reports ranges of $3.5 to $12 and $0.1 to $2.4 per person, respectively. These compare against the international benchmarks in contrasting ways for CapEx and OpEx as compared below.

Table 11.3 Descriptive statistics on major cost categories (whole sample)

		CapEx hardware	CapEx software	OpEx	OpEx support	CapManEx hardware	CapManEx software
Programmes (*n*)		20	20	20	20	20	20
Mean		$207	$25	$8.75	$2.35	$7.58	$0.19
Std. deviation		214.30	51.88	7.85	5.21	10.88	0.68
Percentiles	25	$93	$0.72	$3.56	$0.14	$0.38	$0.00
	50	$164	$2.83	$5.72	$0.51	$4.75	$0.00
	75	$233	$17	$12	$2.40	$11	$0.00

First, comparing CapEx costs, the new data reported here are significantly higher than compared to the WASHCost datasets (Burr and Fonseca, 2013). Based on the consolidated global dataset, the WASHCost research reported the following IQR for CapEx of $31 to $132 per person for rural piped schemes (ibid.). Yet when focusing on just the WASHCost India data the benchmarks are even lower at $23 to $81 for CapEx (Burr, 2015). This means the lower IQR for CapEx hardware from this study is already higher than the highest benchmark reported from the WASHCost Andhra Pradesh research. Second, for OpEx costs, the data from the international WASHCost IQRs are $0.5 to $5.3 and for OpEx Enabling Support they are $1.1 to $3.2 (Burr and Fonseca, 2013). The data from this research provides a much higher range of OpEx costs but lower OpEx enabling support costs. This is considered to reflect the way operational support is structured in India, which includes significant direct subsidy to community service providers during the OpEx stage, which in turn allows the level of OpEx Enabling Support to be lower.

Overall, the whole sample findings from this research indicate higher combined OpEx and OpEx Enabling Support costs than the international benchmarks. Again, when comparing directly to the WASHCost India data the differences are even greater. That study found recurrent costs in the Indian sector hard to come by but reports a benchmark for the combined costs of between $0.2 and $2.5 per person per year (Burr and Fonseca, 2013). When comparing that against the findings from this study, again the lower benchmarks from this research are higher than the upper benchmarks provided by the WASHCost project. The final cost categories reported on are for CapManEx in hardware and software. There were 18 case studies in which CapManEx was reported but two case studies (Odisha and Karnataka) reported no CapManEx hardware data. There were only four cases where any form of CapManEx software was conducted, suggesting this is an area of underinvestment. The data provided gives a range of $0.4 to $11 for CapManEx compared to the WASHCost India range of $0.0 to $1.2. At this aggregate level, the data presented here clearly shows that the levels of expenditure found in these 'reportedly successful' community-managed rural water service programmes are at a significantly higher level than has been previously reported.

Financial cost sharing between support and community

A core contribution this study makes to the literature is to reassess what cost-sharing arrangements can support successful community management in the context of the various calls for a shift to community management plus (Baumann, 2006; Lockwood, 2002, 2004; Moriarty et al., 2013). This section focuses on this issue through analysing the financial cost-sharing arrangements between support agencies and communities. As shown in Table 11.4, the mean proportional contribution of communities to capital costs is 5 per cent, which is around half of the often held standard for 'demand-responsive' community management which was 10 per cent of capital costs in the Indian context (Government of India, 2003; Hutchings et al., 2016). The range given for community contribution to CapEx

Table 11.4 CapEx cost sharing

Descriptive statistics	CapEx hardware support	CapEx software support	CapEx community	OpEx direct support	OpEx enabling support	OpEx community	CapManEx support hardware	CapManEx support software	CapManEx community
Mean	84%	11%	5%	26%	21%	53%	82%	3%	15%
IQR	99–87%	1–7%	0–7%	1–30%	6–18%	52–93%	79–89%	0–0%	11–21%

across the cases is 0–7 per cent. Significantly, there are four case studies where communities do not contribute any money to capital costs which reflects a complete departure from the principle that communities need to make a financial contribution in the implementation phase of a community management programme in order to demonstrate some degree of 'ownership' and therefore commitment to future management. For CapEx software this research indicates a range of 1–7 per cent of total capital costs which is lower than the 10 per cent estimate used by the water and sanitation programme of the World Bank in their calculation of investment needs in the sector (Hutton and Varughese, 2016).

The recurrent cost-sharing arrangements show greater variability in the pattern of cost sharing for recurrent costs than capital costs. Focusing on the operational costs, the IQR for community contributions is 52–93 per cent of the total. Reflecting this finding back against the principles of the demand-responsive approach to community management it is clear that generally speaking communities are not covering 100 per cent of operational costs as is implied under a demand-responsive model (Government of India, 2003; Hutchings et al., 2016). Instead, they are reliant on significant direct and indirect financial support. The direct support through subsidies to OpEx range from 1–30 per cent and the OpEx Enabling Support range from 6–18 per cent of the overall operational costs. In Chapter 2 the financing used in the Capital Maintenance Phase of the service delivery cycle was identified as the critical failure point in many rural water service programmes. The data here shows communities can contribute 11–21 per cent of these costs in successful programmes with the remaining 29–89 per cent covered by supporting agencies. The study found very little CapManEx on software across the cases indicating this is an area of systemic underinvestment, although also perhaps a function of studying generally longer-life piped water supply systems (as compared to handpumps).

The cost-sharing data in terms of dollars invested by support and community is now considered. This helps to quantify the level of financial investment found and shows an IQR of $2.4 to $6.6 per person for community contributions to OpEx costs while the external support for that category is $0.6 to $4.8 each year, as displayed by Table 11.5. This indicates there is a 'willingness to pay' for services through the community management model but tariffs continue to require significant subsidy even to cover the basic OpEx costs. Similarly the data shows that for many of the case studies community members demonstrated a willingness to pay

Table 11.5 Financial data on cost sharing

	CapEx support	CapEx community	OpEx support and OpEx enabling support	OpEx community	CapManEx support	CapManEx community
Mean	$220	$13	$5.3	$5.8	$5.6	$2.2
IQR	$98–245	$0–17	$0.6–4.8	$2.4–6.6	$0.0–8.2	$0.1–4.6

for capital costs of up to around $17 per person for a one-off payment for CapEx and just over $2 annually to cover CapManEx.

Financial costs by sub-group

This section investigates the cost-sharing arrangements by organisational types, technical modes of supply and economic wealth. As the sample becomes stratified the number of cases in each sub-group becomes smaller, yet the approach is still to present the IQRs to indicate that there is variability even between sub-groups. However, it is acknowledged for some analysis this means that the IQR can reflect the entire range of results.

Organisational types

Developing the arguments made in the previous chapter about the different organisational types, this section explains the cost-sharing patterns between these groups. At the enabling support environment level, a notable difference is the higher levels of investment in the service delivery stage in the decentralised LSG cases compared to the other enabling support environment types. This includes higher tariff payments by communities and also greater external support from governments. The interquartile ranges of the other cost categories did not suggest a strong pattern of difference between the models, as shown in Table 11.6.

Table 11.6 Financial costs by enabling support environment type

ESE	CapEx support	CapEx community	OpEx support and OpEx enabling support	OpEx community
SRWSA	$81–208	$0–11	$1.0–4.4	$1.7–3.7
LSG	$115–245	$6–32	$4.4–25.5	$3.4–13.5
Hybrid	$166–331	$0–16	$0.2–5.7	$3.6–13.9
External agency	$37–245	$1–34	$0.4–3.0	$0.4–5.9

Looking at CapEx in percentage terms, the SRWSA and hybrid support approaches have an interquartile range 0–9 per cent and 0–6 per cent respectively. The lower quartile of 0 per cent is considered to be a relevant finding to highlight as it indicates the shift away from communities necessarily contributing towards CapEx, which had been a key principle in the popular demand-responsive mode of community management (Joshi, 2003; Moriarty et al., 2013; Schouten and Moriarty, 2003). With local self-government and external agency support systems there is an interquartile range of community contribution to CapEx of 2–17 per cent and 3–18 per cent, respectively. The high variety indicates the level of investment is programme-specific and can range from significantly below, to nearly double, the 10 per cent contribution level highlighted in the literature as standard practice (Joshi, 2003; Moriarty et al., 2013; Schouten and Moriarty, 2003). The proportional community contribution to OpEx also has high variety across the enabling support environment types. Yet in considering the interquartile ranges, the difference between the two government support programmes types – the SWRSA and LSG – and the hybrid and external agency is with regards to the upper quartiles. Those government enabling support environment types share an upper quartile of 75 per cent while in the other categories it is 90 per cent (external agency) and 97 per cent (hybrid). Again, this is considered indicative of the shift in Government of India supported community management away from the cost-sharing principles that have come to be associated with community management (Joshi, 2003; Moriarty et al., 2013; Schouten and Moriarty, 2003), including that communities cover 100 per cent of OpEx. It is the community management programmes either led by or in partnership with non-government partners where the upper quartiles get close to the 100 per cent level for community contribution associated with the demand-responsive approach to community management (Joshi, 2003). This difference between the domestic interpretation of the community management model and the more international ideals of community management is considered in more detail in Chapter 14.

When cutting the case studies by community service provider types, the previous chapter demonstrated that the registered societies as VWSC model for the community service provider was the typology most associated with good service-level outcomes. The financing of this type of model shows it has the highest level

Table 11.7 Financial costs by community service provider type

CSP	CapEx support	CapEx community	OpEx support and OpEx enabling support	OpEx community
Autonomous VWSC (single case)	$83	$11	$1.0	$5.0
Representative VWSC	$97–257	$0–9	$2.1–17.0	$2.4–6.6
Registered society	$145–238	$11–33	$1.7–4.8	$2.6–13.7
Unregistered society	$37–331	$0–11	$0.2–0.8	$0.4–5.7

of community contribution to CapEx and OpEx suggesting a link between the level of community contribution and the success of these services. Direct support for OpEx and OpEx enabling support are also higher in in the registered societies than the next most successful typology of representative VWSCs.

In considering the proportional cost sharing between community and support, the representative VWSCs – which are so closely intertwined with the local self-government system – and unregistered societies, have very low levels of community contribution to CapEx. The interquartile range is 0–5 per cent and 0–3 per cent, respectively. As the representative VWSC case studies are the institutional set-up most closely linked to the prescriptions of the NRDWP guidelines (Government of India, 2013a), it shows how this shift is part of the institutionalised move away from community contribution to CapEx as a precondition for a community management programme in India. Interestingly, though, the registered society which has the best overall service-level outcomes retains the more conventional cost-sharing arrangements for CapEx with an interquartile range of 7 to 15 per cent for the case studies in that category. This would suggest that there remains a link between community contribution and service-level performance, which can be considered an implicit assumption underpinning the community management model. Comparing the community service provider typologies by OpEx contribution again indicates the lower levels of contribution found within the government-prescribed model of the representative VWSC as compared to the other types of service provider arrangements. That typology has an interquartile range of 21–61 per cent against 36–84 per cent (registered society) and 34–97 per cent (unregistered society) with the single autonomous VWSC having 83 per cent community contribution for OpEx. This is again considered evidence of a distinction between the most recent Government of India influenced case studies, and those that have non-government actors involved as well.

Technical mode of supply

The difference between costs of services by technical system design are now considered. Assessing costs in this way shows that CapEx increases along what can be described as a crude technical spectrum of complexity from the lowest level with the most simplistic technology to the highest level with very sophisticated systems. For example, the gravity-fed MVS in the Maharashtra case study is over 26 times more expensive per person than the handpumps in West Bengal, mainly because of the hydro-geological need for long-term bulk water storage that justifies the construction of a multi-use dam. The other forms of piped water supply regularly have a CapEx in the range of $100 to $200 per person, although the gravity-fed schemes tend to have a higher upper value, which is in part explained by the mountainous contexts of Sikkim and Himachal Pradesh. Here, the technical design of schemes is often more sophisticated while the construction and material costs tend to be higher. This is all compounded by small village sizes that reduce the economies of scale at which work can be undertaken. Higher CapEx costs do not necessarily mean higher recurrent costs though as the Maharashtra case has relatively low recurrent costs partly due to the lack of energy needed for pumping water (the design in this case taking advantage of the higher elevation of the reservoir and treatment works), which is normally a major recurrent cost requirement. In this sense, higher CapEx can be used to reduce recurrent costs but overall the recurrent costs of piped supply are still higher than handpumps.

Costs of water supply in different contexts

This section now considers the differences when cutting the cases by income group. Figure 11.6 displays the data for operational costs, which combines both OpEx and OpEx Enabling Support. It shows the median level of per capita investment in support rises across the economic groupings from less than $1 in the low-income states to around $2.50 in the middle-income states and up to over $4.50 in the high-income states. The biggest difference in the figures is the higher levels of community contribution found in the middle-income states, with a median of nearly $14 per person, compared with $3 to $3.50 per person in the low-income and high-income states. These patterns of cost sharing are considered to be partly reflective of the different types of organisational arrangements across states but also partly reflective of the modes of political economy that dominate development processes in the different income groups. The case studies from the middle-income states can be characterised as being more likely to follow the bottom-up – 'social-democratic' – approach to rural water services in which populations are the drivers of the development process and therefore are willing to contribute financially to public services (as shown in Figure 11.6). However, in the high-income states the approach is more top-down – 'developmental' – as the support entities take on a higher level of investment and communities invest less of their own finance. In the low-income states the overall capacity for investment is lower and therefore this leads to lower levels of investment across the categories.

Costs and performance categories

This section now considers how the level of financing directly relates to service-level performance. It does this first by assessing the cases by the service-level performance category set out earlier in the chapter before a simple correlation analysis between financial categories and service-level outcomes is conducted. As shown in Table 11.8, the high performance category is shown to have higher levels of community contribution to CapEx and OpEx than the other categories – for example, a mean of $10 per person for OpEx in the high category against $4 and $3 in the medium and low categories, respectively. This again suggests a link between the performance of services and a willingness to pay although it is not possible to identify whether higher community payments lead to better performance or is because of it. The low performance category has higher levels of CapEx from the enabling support environment but recurrent levels of investment are generally lower. This would therefore suggest that the level of CapEx is a less important category for determining the level of success.

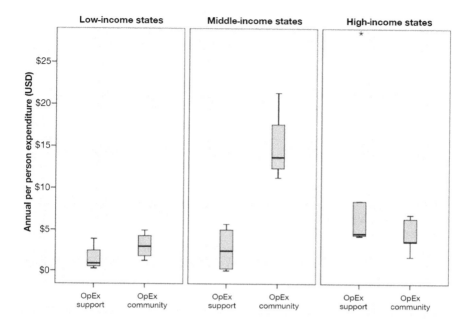

Figure 11.6 Recurrent costs across socio-economic groupings of cases

A correlation analysis was also conducted on a dataset containing all the case studies. The two dependent variables were the median service-level outcomes as per the composite service-level indicator and the percentage of the surveyed population that meets 'basic' or above on that indicator. The analysis shows that there is

Table 11.8 Costs of services by performance category

Performance category	Statistic	CapEx support	CapEx community	OpEx support and OpEx enabling support	OpEx community	CapManEx support	CapManEx community
High	Mean	$208	$19	$2.5	$10.3	$6.4	$2.8
	IQR	$153–231	$11–31	$0.2–4.6	$3.7–13.9	$0–11.3	$0.6–5.9
Medium	Mean	$237	$8	$7.6	$4.2	$3.8	$1.4
	IQR	$67–245	$0–13	$3–8.5	$2.3–6.4	$0.1–6.7	$0–2.5
Low	Mean	$213	$11	$5.3	$3.7	$6.7	$2.4
	IQR	$83–268	$0–14	$0.5–5.1	$1.1–5	$0–2	$0.1–4.8

only one cost category that has a highly significant positive correlation with these two variables. That is the level of OpEx costs. The correlation between the OpEx is medium to large for the Kendall's tau test (Field, 2013). In considering this relationship the relative correlation of the community and support contribution to OpEx was also tested with this showing a significantly higher association between higher levels of community contribution to OpEx and higher service-level outcomes. This again indicates that high levels of community contribution are a good indicator of success but as explained earlier in this chapter it is not possible to answer conclusively whether the higher contribution from community drives stronger performance or whether when good services are provided populations are more likely to pay for them.

Beyond the OpEx category there is no linear relationship that is statistically significant between the cost categories and service-level outcomes for CapEx, OpEx enabling support or CapManEx investments. This would suggest that as the sector increases levels of investment concentrating more financial subsidy directly to service provision tasks is likely to lead to better outcomes. However, the absorptive capacity of the service provider – in terms of being able to effectively use the additional financial resources effectively – is likely to be conditional on broader factors which can most likely be improved by broader investments in terms such as OpEx enabling support. Ultimately, the research continues to suggest that there is no single factor that leads to successful community management. However this section has helped identify the conditions in which success arises within the Indian context which do involve substantial on-going financial support to communities.

Discussion and conclusions

This chapter has provided a descriptive analysis of service-level outcomes and financial data across the case studies. For the service-level outcomes, the analysis shows that community management can deliver high service levels with six cases being labelled as high-performance case studies. Yet even though all cases have 100 per cent access to an improved water source there remain seven cases with basic performance and seven cases that fail to meet government norms. The research therefore adds to the evidence about the limited nature of using the concept of *access* as a measure of success in rural water services (Clasen, 2012). The research also reinforces the importance of having household connections as compared to communal water points, such as public stand-posts or handpumps, with the service levels from these water sources significantly higher than any other type of improved water source.

Beyond the service-level analysis, this chapter provided a descriptive analysis of the financial data to answer research question two on the indicative financial costs of supporting successful community management programmes in India. Table 11.9 provides a summary of the total levels of financial investment in CapEx and recurrent expenditure for different levels of service-level performance from this research against the WASHCost benchmarks. It shows that the levels of investment found

Table 11.9 Summary of financial requirements for successful community management

Source	CapEx (IQR)	Recurrent (IQR)
High service-level performance (Community Water Plus)	$184–247	$6–32
Medium service-level performance (Community Water Plus)	$73–279	$6–28
Low service-level performance (Community Water Plus)	$93–281	$5–13
WASHCost international benchmark (Burr and Fonseca, 2013)	$31–132	$4–16
WASHCost Andhra Pradesh benchmark (Burr, 2015)	$38–66	$0.5–1.6

are higher than these most widely used benchmarks for rural water service financing (Burr and Fonseca, 2013; Hutton and Varughese, 2016). The data from this research is most strikingly different to the results from the WASHCost India study that suggested extremely low levels of financing in the Indian sector, especially for recurrent costs. This research indicates an inverse situation in which the Indian sector has higher levels of financing than the suggested ranges for low- and lower middle-income countries.

Conducting this research revealed many of the costs that go into the Indian sector are opaque and hard to trace without a detailed case study approach to the research. In a government programme, for example, finance comes through various state and federal government funds directly into the local self-government Gram Panchayat accounts while other funding can be funnelled directly to VWSC accounts or may come from indirect subsidies – often unacknowledged – such as discounted rates for electricity power. For this reason, it is contended that previous research has underestimated the cost of delivering services in India. The WASHCost project itself acknowledged that evidence of recurrent costs were hard to come by in the Andhra Pradesh study (Burr, 2015) and therefore many estimates were used. In this sense, although the research cannot be considered representative of the general workings of the Indian sector, the level of divergence between the estimates provided and the recurrent estimates from the WASHCost project, suggest that the accepted understandings of working in 'low-cost' India should be revised as there is significant financing being poured into the Indian sector.

In terms of the financing patterns most strongly associated with delivering good service-level outcomes, the analysis presented showed that higher levels of OpEx correlated with high performance outcomes. It was shown that the level of community contribution to OpEx was especially strongly correlated. This indicates that although the sector is concentrating on how to structure and finance an effective enabling support environment for community management, providing direct financing to service providers and promoting community tariff contributions can be considered useful approaches for driving up performance levels. Beyond those trends, another key findings was in terms of the fragmentation costs between different agencies within single programmes. This shows that when assessing the costs of rural water supply it is important to consider both direct costs incurred by service providers but also enabling support costs, both direct and indirect, incurred by the various entities operating at the enabling support environment level. The broader implications of these findings both in terms of policy, practice and theory are revisited in the final chapter of the book.

12 Monitoring and regulation of community management

This chapter now focuses on the specific challenge of monitoring and regulation of community management. Monitoring drinking water service delivery in a country as large as India not only poses challenges in data collection but in the analysis and interpretation as well. However, unless the services so expensively developed are being monitored, with some form of regulation in place to influence quality and consumer contributions, it is difficult to check the functionality or ensure the sustainability of water supply. For a vast country like India, with its approximately 600,000 villages (Ministry of Drinking Water and Sanitation, 2016), designing a common monitoring programme is not only difficult but making it functional, and used, poses many challenges. It has been understood from the research that there is a tendency for even the most successful programme villages to 'slip' back to inadequate service if continuous monitoring, and subsequent support, is not ensured. The research described in this book has focused on systems which have been in operation for at least five years, to ensure that the measure of 'success' includes at least a reasonable period of functioning. However, there is concern that systems which had been researched as success stories in previous generations (though researched for different purposes) have not continued in a positive way over the ten to twenty-year time scale. 'Slippage' is the term used in India to describe the way in which completed projects, initially delivering the required standard of water, subsequently regress and fail; figures explained in detail later in the chapter report 59 per cent of water assets tested in one state are failing to deliver potable water.

The particular focus of this book is the interaction of communities managing their water services in partnership with their enabling support environment, which is usually the state in its various guises. We are therefore concerned in this chapter as to the role of communities and their support environment in monitoring and regulating. Communities who manage, or who are involved in the management of, their water supply service need to know how well their system is performing. They need to know how effectively it is delivering clean water (indeed, if that water is actually potable); whether all in the community are accessing a fair share of that water at a fair price; how viable their service is in financial terms, including the ability to finance future renewal of fixed assets; and perhaps the extent to which their way of managing water is efficient, that is could more be done at a lower cost to the community? Communities need this information about their own services in order to make valid decisions as to future management. They also need some of this information about their neighbouring communities in order to have some comparison by which they can reflect on their own performance. Such management information, 'key performance indicators', can then be used to reflect against national performance goals and standards, an important tool when significant national and state funding is being used to support community services.

It can then be asked what communities can do to measure their own performance, remembering the old adage that 'if it isn't being measured, it isn't being managed' – in the sense that communities need to measure for themselves in order to be able to manage themselves. In addition, there are questions as to what level of external support communities need to be able to monitor performance indicators appropriately, in access to water quality testing laboratories for example, as well as what level of comparator information can be made available by the enabling support environment. With that 'support environment' funding over 90 per cent of capital expenditure and, on average in this research's 20 case studies, approximately 50 per cent of recurrent expenditure, the external support to community management should be taking some view on the level of performance achieved at what overall, and community level, cost. This can be seen as a form of external 'quality regulation', the support environment ensuring that water quality and equitable access are being delivered, while also considering 'economic regulation', whether consumers are paying an appropriate price for this service, recognising that subsidies from general taxation are a critical aspect of ensuring services for all.

Similarly, we need to investigate what level of external supporting influence or 'regulating' is acceptable over community tariff-setting, and subsequent payments, that deliver an appropriate contribution that allows for and promotes actual community management in addition to a community level of understanding of the value of water as a scarce resource. There is therefore a tension between community empowerment and the need for some external monitoring and regulation, taking into account the degree of external support being delivered. And as in all monitoring and regulatory systems there is also the challenge of proportionality. There is an apparently insatiable desire of support entities (and researchers) to have access to ever more, and ever more detailed, performance information to support monitoring and regulation which may, in the end, entail a disproportionate cost in collecting that information relative to the possible benefits of its use, indeed, if it is ever used at all. This chapter looks at the approaches to monitoring and regulation across the 20 case studies, with a particular focus on states where innovative systems are being used; in India this means the use of information technology and, increasingly, smartphones.

Community-level monitoring

The research found that communities monitor what is most important from their perspective, that is are they receiving water and if not, why not? At community level this tends to be an informal level of monitoring with few, if any, records kept. Problems are solved as and when they arise which means that without performance indicators community management tends to be reactive. Only if the issue has to be escalated to the support entity does it become a matter of record, particularly in states which have grievance or complaints systems. Where individual consumers' supply is not directly affected, by pipe leaks along the road for example, one community manager explained that now the community is self-monitoring they are reporting such leaks of 'their' water to their local committee, whereas in the past they would have ignored the losses of 'government' water. But there were rarely records of the leaks and therefore no patterns to discern to aid decisions as to whether lengths of pipes needed to be replaced, rather than simply on-going fixing of a specific problem.

The second major concern of communities is finance, that is who is paying their water tariffs (noting the positive support reported for those in temporary payment difficulties) and the extent to which their locally incurred costs, employment of a pump operator and an electricity bill for example, are covered by their monthly income. This information is made available to the management committee, sometimes directly to the community through 'transparency boards' (the January 2016 board in Figure 12.1 from our Punjab case study detailing monthly income, expenditure and resulting savings had clearly been updated since the initial visit in October 2014) and is usually looked at also by the junior engineer in the supporting rural water supply agency. Those community service providers which are 'registered societies' are required to have their accounts externally audited annually, a source of some satisfaction to those reporting the results.

Water quality testing is undertaken with samples sent to government laboratories for testing (often free of charge), sometimes monthly, sometimes annually, with the focus on the chemical attributes and contamination with little testing for biological contamination. Such monitoring can help communities understand how important it is to make decisions at the right time. Water quality monitoring in one of the Kerala case studies has caused the community to seek additional external funding to support a new water treatment works. In Meghalaya, as another example, the village water committee in Ompling has invested in an extra motor using the savings generated from their water tariffs. This not only reduced the burden on the existing motor but helped in preventing the machine to be overloaded, with likely consequential repairs. Currently both the motors share the load and the supply is continuous. This decision could be implemented easily as the community had ringfenced their funds and continuously monitored the fund flows into their water committee account. Further the technical engineer present in the water committee helped to retrofit the old system and adjust the capacities so that both pumps worked alternately improving the efficiency of the system.

Similar examples of the benefits of technical expertise within community management committees were found in Kerala and Punjab as well. Retired engineers from the state engineering department were able to assist their committees to determine feasible options resulting from monitoring. The majority of committees are not so fortunate and require their enabling support environment to help

Figure 12.1 Updated transparency board for community monitoring, Punjab

them interpret necessary action resulting from monitoring. The ever-increasing use of information technology generally is now being harnessed to support rural water. In Odisha's community-managed systems, those supported by NGO Gram Vikas in this research's case studies, water pumps can now be operated remotely via mobile phones, presumably also recording pumping hours and volumes, along with automated chlorine dispenser disinfection units, giving communities considerably more understanding and control over their own services. Such an approach is likely to be copied by many other service providers in the near future.

Enabling support environment – state-level monitoring and regulation

Water being a state-level responsibility in India, each state has its own variations on monitoring the performance of community service providers and, from the perspective of the support entities, the extent to which state and GoI funding is delivering the political imperative of 'fully covered' habitations. The first concern is therefore whether the budgeted funds are being spent in a timely fashion on new works which will increase service coverage. The second major driver appears to be the water quality information and other data needed for the annual return to the national government's Integrated Management Information System (IMIS), described in more detail in the following section. Subsequent to those concerns each state has variations on the theme of a monitoring database, usually on a spectrum of sophistication which reflects the state's view of information technology as a driver of economic growth. We follow the example of Andhra Pradesh, below, which developed an early and impressive example of software-based monitoring, now being taken onwards in the newly bifurcated state by a chief minister committed to using information technology to all possible extents. As a subset of state-level monitoring, we then explore the Punjab example of a sophisticated state-wide complaints and grievance resolution system which perhaps empowers consumers rather more than it supports community management.

State-level monitoring in Andhra Pradesh (now Telangana and Andhra Pradesh)

Before the recent bifurcation of the state, the precursor Andhra Pradesh Rural Water Supply Department (RWS) developed a sophisticated database and management information system, 'WaterSoft', to provide the necessary understanding of location, quantity and quality of water supply for the 72,000 habitations in the then state. The database was developed by the National Informatics Centre (NIC) and had been operational since 2008. Modules were developed incrementally, based on requests from the RWS, with a total of six finally in use: works monitoring, assets monitoring, drinking water status, drinking water to schools, sanitation status and sanitation status in schools. The works monitoring module requires assistant engineers to update every month the progress of new works, that is augmentation of or new multi-village or single-village schemes. This module logged progress against milestones and also the value of budgeted expenditure, though not the details of actual expenditure. Information had to be entered prior to sending bills for payment.

The assets module replaced the original paper registers and shows assets created by RWS in each habitation. Location-wise entries are updated every year showing the status of assets as on 1 April (spill-over works are carried over into the next year). The asset register included the status of piped water schemes (PWSs) such as location, year of construction, working condition and cost of construction linked to a unique habitation code along with water source details: location, source condition, average static water level (metres), yield (litres per minute), source depth, alternative water source type and distance (km) and water quality. The detail recorded was extensive, also including pump details (location, make of the pump and its capacity (HP)), year of commissioning, hours of running per day; usage of regular and stand-by pumps, design head (metres) and flow (litres per day); raw water collection details, such as capacity (litres), diameter (metres), depth (metres) for surface water schemes; summer storage tank details, if any; filter details, if applicable: including slow and rapid sand filters, micro filters and reverse osmosis (RO) plants.

The assets module detailed operations and maintenance (O&M) issues regarding public water supplies such as: the responsible agency, annual cost of power, consumables, wages and other local costs, the monthly tariff structure and tariff collection per year, source of O&M funds (e.g. Gram Panchayat from the Twelfth Finance Commission funds set aside for RWS O&M expenses) – although no details are available with the RWS since those are with individual Gram Panchayats. There are further modules for additional information to be recorded. The executive engineer of every RWS division has a password for data entry, and is the only official authorised to make changes in the entered data, but these have to be based on hard-copy changes verified by lower staff such as the assistant engineer or section officer.

Interviews with users of WaterSoft as part of this research suggests the system has been useful for financial management, giving a degree of transparency in payments to contractors for new schemes. The other main focus on fixed assets has limited value as information is not being promptly updated and users feel there is always a discrepancy on data. Similarly, the updating of information on habitations is not undertaken regularly so users find that there are a lot of discrepancies in the data. Overall there was no streamlining or verification of data systematically, though occasionally there has been some third-party assessment done by national government.

The results of water quality monitoring are entered into the system and data can be retrieved for the last three years but it is reported that it may be difficult to get three years' data for the same source. Therefore, there is little opportunity for professional, or community, oversight of long-term water source quality trends. An additional facility provided a toll-free number for complaints or grievance monitoring but again there are other reporting formats, and databases, during droughts and for summer water shortages to meet the varying urgent political needs.

Overall there is a sense that the system has not been used to its potential by communities (or local government Gram Panchayats), by field-level staff – apart from the new works accounting processes – or by state planners, apart that is from providing figures for overall coverage. There is no sense that the asset management information was being used to inform policy-makers and budget planners about the long-term funding needs for capital maintenance. Similarly, the system could not directly meet the needs of national government (see following text) as there were no links between the national ministry's IMIS system and the state WaterSoft database. IMIS requires an annual update from the states in the month of April/May with no subsequent updating after that. Data have to be prepared specially by district and state officials for the annual submission to IMIS – with similar challenges being reported by other states.

On a more positive note it is reported that state bifurcation did not cause major problems, as the contractor, National Informatics Centre, carefully separated the data according to district. There were reportedly problems with three *mandals* in one area and they had to be added manually to all the databases. Although the National Informatics Centre was able to implement WaterSoft in some other states, the performance of the system shows a mixed picture of effectiveness. At the time of the fieldwork, the WaterSoft system was being used mainly to generate the information required by the principal secretary for guiding investment and other decisions at the very top of the hierarchical chain of command. As explained earlier, executive engineers and assistant executive engineers (AEEs) at the field level continue to rely on hard-copy files and measurement books for day-to-day operations and hardly ever use the software to guide their own decision-making. At all engineering levels, from the AEE to the chief engineers at the headquarters in Hyderabad, the WaterSoft system is basically regarded as software that has to be populated regularly for information that is not of direct use to them, but is required at the highest level of decision-making within the department.

With no apparent use of the system by communities, or their Gram Panchayats acting as community water service providers, and little use by their enabling support environment professionals in preparing for the next generation of challenges – capital maintenance along with system enhancement and expansion – WaterSoft gives the impression of being an underused while overly complex and sophisticated approach to the problems facing water providers in what are now the two states of Telangana and Andhra Pradesh. This is the challenge of so many monitoring approaches for rural water.

Telangana continues to employ WaterSoft with little enthusiasm or use. The new state of Andhra Pradesh now has WAMS – 'Water Asset Management System', a development of the WaterSoft approach by NIC, which appears to have taken past criticisms on board and, among various smartphone-based app facilities, also has an overall 'Dashboard'. At the time of writing (WAMS, 2016) reports that of the total assets monitored 59 per cent were failing to deliver potable water, of the protected water supply schemes 13 per cent of the 62,928 schemes were not working, of the total assets listed (280,894) 28 per cent were not working, which included 170,629 handpumps, of which one-third were not working and 20,359 open wells, of which 48 per cent were considered to be not working. In these early days for the new state it can only be assumed that such explicit and public monitoring information will be used by communities and planners to good effect.

Complaints redressal system – Punjab

As a subset of monitoring, perhaps more of a short cut to action bypassing long lead-time monitoring, the water sector in India has a strong history of complaints recording. Junior engineers are expected to respond, with varying degrees of alacrity, to the information in dusty complaints register books – these records often being the only performance indicators available at local level. The Punjab water supply and sanitation department (WSSD) has set up a centralised public complaints system for community rural water supply in the state which both monitors performance, through the type of complaints being received, in addition to monitoring support environment response to those complaints.

The 'Shikayat Nivaran Kendra' (SNK) system, based at Mohali, allows all rural water consumers to lodge their complaints through a toll-free number, as well as retrieve the latest information about the status of the complaints lodged by them. The call centre is reportedly operated around the clock, throughout the year. It uses an 'Interactive Voice Response' system which helps customers to retrieve the required information, lodge complaints and place other customer service requests at the call centre.

At the time of registration, the name of the appropriate WSSD engineer relating to the complaint village is shown on the screen (where using the smartphone or web-based app) and the complaint is forwarded to them through SMS and email 'for immediate redressal'. Information regarding the status of various complaints registered at SNK can then be monitored by department officials. The time allowed for a response to complaints has been pre-determined for typical issues – the designated officers are expected to solve the complaint within this fixed time frame and report back to the Centre through their own text message so as to enable SNK to inform the complainant as to progress. Where the case complaint is not rectified within the stipulated period, it is forwarded to the next level official for their intervention after every 48 hours, cascading upwards through the hierarchy to superintending engineers and ultimately the chief engineer and the assistant secretary. Meanwhile consumers can check the status of their complaints through the unique complaint number provided to them.

As described, all the complaints have to be addressed in a certain time period and officials monitor the status of the complaint on a daily basis. Typical issues raised as complaints in a recent summary shared with the researchers are 'failure of water supply due to absence of operator', 8 per cent of total complaints, an issue which the responsible junior engineer assigned to that village is expected to sort out within one day but was referred upwards on 6 per cent of occasions in this particular sample due to on–going problems. Other complaints were 'bad quality of water' (9 per cent); 'electrical or mechanical fault' (11 per cent); 'large scale leakages in pipes' (17 per cent) and 'failure of water supply in some specific area: may be due to uneven topography or some other reasons' (45 per cent). This latter challenge, the junior engineer being allowed three days to solve, led to the most referrals to 'intervention of senior official' at 8 per cent of complaints compared to the average of 5 per cent overall.

The monitoring system, available in summary on the WSSD staff mobile phones to complement the extensive database, shows outstanding complaints by district with a colour coding approach for each junior engineer, red for those with more than three outstanding complaints against their name, yellow for two. Staff in the department suggest the benefits of this system are the check on absenteeism among lower ranks of employees in remote villages, the help in achieving higher operation and maintenance standards, the timely delivery of services, departmental performance being assessed on a daily basis with the ultimate benefit that 'availability of clean water leads to good health with huge social sector benefits'.

It might be presumed that this complaints system should only operate as a backup to the community's own management responsibility whereby complainants only escalate their concern when referral to the village water and sanitation committee has not produced results. However there does not appear to be this requirement, the department has promoted this scheme extensively through advertisements in the local print media, in the Punjabi News Channel, with information displayed at prominent places in all the villages by wall signs stencilled in the vernacular language, with an advertisement inserted in 23,000 copies of a newsletter being circulated quarterly in the various villages under the PRWSS project as well as jingles aired on All India Radio.

By these means it appears the department is setting itself up as the responsible service provider rather than as a support entity. A more community-oriented approach would be for members of the management committee to have the right to access the complaints system in order to get any required external support. The way the system now works can be seen to disempower local management.

Monitoring and benchmarking

A good monitoring system would give both the enabling support environment and the community service provider a set of indicators and the rationale for using those indicators. The guidance would need to include the data required for calculating the indicator, the formulae for its determination with some way of understanding the reliability of the measurement.

The Ministry of Urban Development in India has a *Handbook on Service Level Benchmarking* (MOUD, undated), for city-level service monitoring, which is a good example of how a rural water service benchmarking handbook could be prepared. The MOUD handbook not only emphasises the need to 'keep systems simple' (ibid., p. 80) but also suggests 'minimum frequency of measurement of performance indicator' and 'smallest geographical jurisdiction for measurement of performance' along with a 'reliability of measurement scale' ranging from A (highest/preferred level of reliability) through two intermediate levels (B & C) to D 'lowest level of reliability' with a full description for each indicator of the situations of measurement which might lead to performance indicator information being classified at each level.

The Ministry of Rural Water Supply and Sanitation has considered developing its own rural water performance monitoring handbook to give the minimum service-level targets or 'benchmarks' which communities and their support entities should be aiming for, with key performance indicators to be included, over and above the rural norm of 55 lpcd of potable water target (now revised upwards to 70 lpcd). The researchers held workshops in Delhi, Kerala and Odisha for support entity personnel from 15 states and worked through with them the indicators they thought were most needed at state and community level.

These water sector professionals clearly targeted service delivery output indicators of quality, quantity and hours of supply along with technical performance measures of leakage and network pressure (to ensure those at the end of the pipeline receive an equitable service) supplemented by management indicators of complaints, tariffs and finance, and functionality of community management with one outcome measure of water-related disease. The workshops recommended that 'performance report cards' should be developed for easy access to and understanding of the data such that every community service provider prepare a report card as they move towards a more professional and accountable approach to service delivery.

National-level monitoring

National-level monitoring, as indicated earlier, is based on the Ministry of Drinking Water and Sanitation's IMIS system (Ministry of Drinking Water and Sanitation, 2016). This recently developed database monitors the National Rural Drinking Water Programme down to habitation level, that state-level information being separately processed and reported to the Ministry each year through additional data entry, the software systems being unable to abstract information directly from the states' different versions of monitoring software.

This centralised management of data delivers online availability of information which is designed to facilitate generation of Management Information reports 'using dynamic user defined templates' which ideally 'ensures timeliness of progress reporting, improved efficiency in planning and programme monitoring and enhanced transparency'.

IMIS is a comprehensive web-based system that enables the states and the central government to monitor the progress of coverage of habitations, rural schools and Anganwadis (habitation-level mother and child welfare clinics) with drinking water supply and the water quality status of drinking water sources through a common format. The habitation-level data on rural water sources, water supply assets and water recharge structures are also captured in the IMIS along with their longitude and latitude locations. Functionality status of the water supply systems, reasons for non-functionality of systems and source-level water quality information is an integral part of this database. Detailed information on water quality status of all villages is available on the system.

There appears therefore to be considerable information available at the national level and details as to how it is being monitored. However, the quality and consistency of the data in IMIS is questioned by many as the state, district and national-level figures cannot be consistently reconciled.

An important study by Westcoat et al. (2016, p. 1) aims 'to evaluate IMIS as a national rural water supply monitoring platform'. They find that, as for the state-level database the authors investigated, the national database is useful for monitoring programmes aimed at increasing coverage and access but they find little evidence that state or local sector planners can use the information to assist in their own programmes. They similarly find that the structuring of information in the database means that only limited analysis can be undertaken without considerable additional processing of the datasets. Their regression analysis experiments 'showed that it currently has significant limitations for policy analysis' (ibid., p. 16).

Gaps in the existing monitoring

In summary, national-level monitoring policy appears to be quite sophisticated, but in reality represents over-collecting of information which can only be used for a single purpose of providing coverage figures. The information asked for by the national government is different from that required by states and as the districts have to fill in the information in a separate format there is no system of retrieving the data systematically. Monitoring of the current system largely focuses on the new works and on-going physical and financial progress so that both state and national government can proclaim coverage numbers. What is needed is monitoring of the quality of the resulting service delivery. It is essential to measure service levels such as quantity, reliability, quality, accessibility and water availability timing. However, the current system is not geared to monitor these indicators which are so important. Though there is occasional water quality testing of water sources, the on-going water quality data over the period of time for the same source is not available. Water quality data is very important as many new schemes are planned basically to overcome the groundwater chemical issues of fluoride and arsenic as India has already achieved the MDG of accessing water, even if no one knows to what extent it is safe in bacteriological terms.

Further, the water supply monitoring systems are designed in an overly-sophisticated way which limits the ability of communities to play their role. Having got past the hurdle of water source sustainability, the question is asked as to what extent monitoring is supporting the community to find appropriate solutions. And now with the move towards more centralised 'multi-village schemes' it is uncertain if there is any role at all for communities to be involved in monitoring these complex, if conventional in high-income country terms, rural water schemes. The case of Maharashtra, where the state water department takes care of everything and communities only contribute water tariffs (the least community-involved form of 'community management'), has no community functions at all in monitoring beyond some limited reporting of service failure at the village water distribution level. In contrast the effective community management evidenced in the Kerala Nenmeni case study is now accessing significant external resources to introduce sophisticated water treatment processes to deliver potable drinking water. How can that community be supported in monitoring on-going water quality challenges?

It is in the context of these questions that the issue of 'regulation' and regulating needs to be addressed, a subject that is well-understood in the context of urban water supply but that has to be considered with some trepidation for community-managed rural water supply.

From monitoring to regulating

Regulation, in this context, is the art of enabling and requiring, possibly sanctioning, both community service providers, and their enabling support environment, to deliver the goal of affordable clean drinking water for all. If the monitoring process tells us the existing situation, regulating is the external oversight that nudges, and then demands, action to overcome any short-comings and ensures resources are available to address any short-comings.

The approach has been for the state rural water supply entity to act as service provider and monitor, using communities as their agents as policy increasingly dictates, with only marginal oversight from national monitoring of national standards. The idea of regulating is to separate out the function of overseeing the achievement of standards from the role of delivering services. That regulating function also requires oversight of the resources available to achieve the mandated standards. Communities, and their support entities, cannot be held liable for failure to achieve standards if the necessary resources, whether financing through community tariffs or state taxation or qualified staff, are unavailable.

India is famed for its very necessary bureaucracy, how else can 600,000 villages be effectively and efficiently served? But the terminology of the often resulting 'red-tapism', usually used in a pejorative sense, is a warning of what can happen to the art of regulating. Much of the value of community involvement in rural water supply comes from the enabling of communities to move forward with their own local responsibility and accountability, in solving their own challenges quickly and responsively without too much of the 'dead hand' of bureaucracy slowing them down.

The regulatory aspect of 'Community Water Plus' therefore is recommended to be 'light touch' with the emphasis on positive handholding rather than any sanctions for failure, though there may be a need for sanctions to be imposed on the support entities when they are slow to make available the resources needed to overcome any quality challenges. And it might be that a more independent view will become necessary over time as to the level of tariffs which communities are charging themselves. Communities are adept at doing just enough in terms of tariff collection with the remainder expected to come from state-level general taxation. Over time the balance may need to move more towards direct consumer tariffs in order both to empower local community management with additional and directly available resources and to help those communities understand the value of scarce water resources.

Present regulation or oversight comes from, for example, the junior engineer attending the board meeting of water committees every six months, supporting the committee as they look at and analyse what monitoring data exists, and as they consider their financial accounts. The engineer can not only be the advisor to the water committee, but also the enabler and the demander of performance, whether it be water quality or tariff-setting and collection or more efficient operations.

There is already significant on-going regulation of capital expenditure by the use of state-level 'standard schedules of rates'. These schedules define, in impressive levels of bureaucratic detail (see for example Government of Andhra Pradesh, 2015) the amounts which contractors can claim for various labour and machine inputs to new works, even to the level of defining, in this case, labour rates for 104 different categories of worker and contractors' profit and overhead charges to an impressively precise 13.615 per cent. However, this regulation of inputs, while protecting the state against excessive, and possibly corrupt, expenditure on new works loses the opportunity for contractors to be incentivised to do better. Good practice regulation, through its nudging, facilitating and incentivising processes can require contractors and suppliers to deliver innovations in outputs (with some regulators now moving the focus even further to outcomes) with the potential for benefit sharing as a result of reduced costs and improved systems.

Perhaps it is too soon in institutional terms for this level of regulating but since the conclusion of the fieldwork for this research, the draft National Water Framework Bill (Union Ministry of Water Resources, 2016) has been published, which requires that 'All States shall establish an independent statutory Water Regulatory Authority for ensuring equitable access to water for all and its fair pricing depending on the purposes for which water is used' (ibid., §22(2)).

It is likely that such Water Regulatory Authorities will have as their primary focus the challenges of urban and irrigation water supply, but this is a significant development which could easily have unintended consequences for rural community-managed water supply. For, in addition to the tendency for such institutions to become overly bureaucratic, the draft Bill does lay down the premise that 'Water as a part of water for life as defined herein, shall not be denied to anyone on the ground of inability to pay' (ibid., §22(a)) and that 'For domestic water supply, a graded pricing system may be adopted, with full cost recovery pricing for the high-income groups, affordable pricing for middle-income, and a certain quantum of free supply to the poor to be determined by the appropriate Government, or alternatively, a minimal quantum of water may be supplied free to all' (ibid., §22(d)).

The appropriate government, under present rules, is the Gram Panchayat who are unlikely to implement 'full cost recovery pricing' for the high-income groups in the village, the higher-income groups usually being well represented in the local council. However, it is also likely that local government will welcome the idea that 'water may be supplied free to all' while ignoring the caveat. One of the benefits of the national and state-level monitoring systems described earlier, though unused to date, is that there is an increasingly sophisticated level of information on the fixed assets employed in rural water supply. This enables any formal regulator to understand the implications of financing the renewal of those fixed assets, both levels of finance and timing of renewals, and to require state governments to have plans in place to deliver their share of finance.

The draft bill also requires that 'Water charges shall be determined on volumetric basis' (ibid., §22(3)). In our cases, only Kerala, Punjab and Karnataka in one village were using household meters, with WASMO in Gujarat utilising bulk water meters. The additional costs of meter installation and meter reading and regular meter replacement are likely to be a significant additional burden on community water supply and can be seen as an unnecessary level of monitoring at this stage of rural development.

It is likely that in cases such as Maharashtra, where there is no community empowered to function and MJP has taken all responsibilities for service provision in exchange for consumer tariffs, there will be an important role for a Water Regulatory Authority (WRA). The WRA, being separate from the state water board (MJP), will be sufficiently objective that it can guide MJP on how much to charge for O&M while keeping the costs down. Similarly in Tamil Nadu, with major new bulk water schemes such as Hogenakkal, a state regulator might be useful in adjudicating tariffs for that bulk water as it is sold on for village distribution. However, with present charges to the distributor of INR3 per m^3 when regulatory accounting approaches would suggest a cost of INR93 per m^3, getting the balance right might take a new regulator quite a while.

This study has found that successful community management of rural water delivers through consumer tariffs representing approximately half of recurrent costs with all households paying the same amount in any area. It is to be hoped that the sophistication of differentiated tariffs appropriate to urban areas are not so applied to rural water that the result is a reduction in the revenues available to communities to manage their water supply.

Using conventional regulatory accounting approaches to tariff determination, particularly with respect to depreciation charges for renewal of fixed assets, the present average annual tariff/payment per person of PPP $6.0 would need to be at a level of PPP $24.2 to reflect full cost recovery pricing (with a 30-year asset life and 5 per cent nominal cost of capital). The introduction of differential pricing, and the confusion over the amount to be shared between tariffs and taxation for average users, will add another level of complexity.

Conclusions – monitoring and regulation

This chapter has shown that monitoring is taking place across the case studies but only with limited benefits to the consuming communities. The software systems for recording monitoring information have no useful connection between states and the government and while delivering good enough information on progress in implementation of new schemes they cannot yet help in ensuring on-going water quality and asset capital maintenance and renewal.

The introduction of state water regulatory authorities is more likely to lead to bureaucratic, prescriptive regulation by norms rather than incentive-based, light-touch facilitation and enabling of community service provision. It will be up to the Gram Panchayats to maintain the authority of their village water and sanitation committees as the community service provider, looking to the support entity for appropriate oversight.

In the case studies we are seeing the transition from volunteerism through community-based management to '*utilisation*' of rural water services whereby community involvement may well be primarily through paying water tariffs, though we trust there will be some on-going community involvement in oversight even then. That change has a corollary or parallel process in monitoring and regulation, moving over time from handholding and support through the beginnings of guidance on tariff-setting and maintaining quality standards on the basis of community management through to more objective and more independent tariff advice and standards of services delivery monitoring with subsequent holding to account of the provider, beyond complaints management, and even the beginnings of penalties for government service provision which is not delivering adequately.

13 Aspects of gender in community management

This chapter focuses on the gender dimensions of community management. Gender is a concept that refers to socially constructed roles, behaviour, activities and attributes that a particular society considers appropriate and ascribes to men and women. These distinct roles and the relations between them may give rise to gender inequalities where one group is systematically favoured and holds advantages over another, stereotyping the roles and behaviours. Women are most often the users, providers and managers of water in rural households and are the guardians of household hygiene. If a water system breaks down, women, not men, will most likely be the ones affected, for they may have to travel further for water or use other means to meet the household's water and sanitation needs.

Given their long-established, active role, women usually are very knowledgeable about current water sources, their quality and reliability, and any restrictions to their use. Men are usually more concerned with water for irrigation or for livestock. Because of these different roles and incentives, it is important to fully involve both women and men in demand-driven water and sanitation programmes, where communities decide what type of systems they want and are willing to contribute financially. As part of the Community Water Plus project an attempt has been made to understand the key issues of how men and women were involved in the process of planning and implementing the water supply schemes and what key roles and responsibilities they have been playing for sustainable service delivery.

It is essential to mention that when we talk about gender usually the focus goes to the women first as they do not share equal space and recognition in the development sector. Further women get limited space and are usually denied the opportunities; community water management is not new with respect of this discrimination, particularly in rural areas. Men are usually dominant when it involves technology and money management and give less space for women despite their success in other areas. Based on the discussions in the field, and observations during the focus group discussions, gender issues are discussed under the following headings: representation in membership; participation in planning, design and implementation; participation in decision-making; participation in financial management; participation in training and capacity building.

Women's representation in village committee membership

Analysis of the 20 cases across the Community Water Plus project reveals that women are involved in committees according to the government norms; in most of the water committees 30 per cent of the membership is reserved for women according to the 73rd Panchayat Raj amendment. However there are exceptions where there are all-women committees (as in Gujarat) and 50 per cent women representation (as in Gram Vikas) observed in cases promoted by NGOs and government–NGO partnership models.

In Gujarat, WASMO provided women with a platform to voice their issues by making it mandatory to have at least one-third of women members in the Pani Samiti (water committee). In Gandhinagar District, Motipura Veda has 100 per cent women members in Pani Samiti where as Amarpura Kherna has a representation of 96 per cent women in the Pani Samaiti. Sardhav has stuck to 33 per cent reservation and filled in four out of eleven positions with women members. While in Kutch district, Bharasar and Shinay has a representation of about 50 per cent women in Pani Samiti. Kanakpar has encouraged women and set an example by having a 100 per cent women members in the Pani Samiti. These special efforts not only increased women's participation but helped in empowering them to improve the functionality of the water committees. A similar approach has been adopted by Gramvikas in Odisha, in Jharkhand and West Bengal and in World Bank assisted projects in Karnataka and Kerala which have at least 50 per cent women in the water committees, indicating the special efforts made by the organisations.

In Maharashtra, MJP has laid down a mandate to ensure a minimum of 33 per cent women in the water committees. In other cases such as Rajasthan, Chatthisgarh, Sikkim, Madhya Pradesh, Karnataka and Tamil Nadu they have followed the government norms of 30 per cent, however women in these committees are passive observers and do not have much role to play. Most of the time the husbands of the members participate in the meetings while the women sign the minutes of the meeting. In the case of the Reverse Osmosis treated drinking water case study villages supported by Bala Vikasa, there is a participation of 20 per cent and 40 per cent of women in the water committees of Pedapapaiahpalli and Koppur respectively. The Naandi Foundation RO treated water model does not have a water committee but in Atkuru village the operator is a woman. In the SWN-MARI case there are no women representatives in the committees.

Though women's empowerment is high in Kerala, the Kodur example showcases less participation from women. The Keriparambu scheme has 25 per cent representation of women whereas Peringottupalem has only 9.5 per cent. Cheruparambu has no women in its committee. The situation is different in the World Bank assisted Nenmeni water committee where 50 per cent of the committee are women and are actively involved in the management. They are part of the Executive Committee (five of the nine members are women) and are members in various zonal committees too. They have clarity of their roles and responsibilities and besides participating in the meetings, they discuss administrative as well as scheme-related issues.

Table 13.1 Women membership of village water and sanitation committees

Region	Percentage of women in VWSCs and comments	
NEO-PATRIMONIAL		
Chatthisgarh	30%	Very nominal role
Jharkhand	50%	But special programme of training of women as jalsahyyas (water mechanics)
Madhya Pradesh	30%	Nominal role despite the mobilisation programme taken up by the NGO
Odisha	50%	Intentionally 50 per cent women are made members and the active members are made secretaries of the water committees
Rajasthan	0–30%	There is no data on women in committees for two villages, however they are supposed to follow the government norms
SOCIAL DEMOCRATIC		
Karnataka	30–50%	Women do participate in meetings though do not take key decisions but they're aware of all the activities
Kerala (Kodur)	0–20%	Women are active in other Panchayat activities but there are no women in the water committee
Kerala (Nenmeni)	50%	Women play an active role, are informed about all the decisions and when needed approach the officials for technical support
Punjab	30%	Men take active role while women are aware and informed about all the decisions
Telangana/ Andhra Pradesh	20%	Women's role is limited and largely maintained by the men. However women technicians are trained
West Bengal	60%	Women play an important role in managing the handpumps
DEVELOPMENTAL		
Gujarat (Gandhinagar)	80–100%	Women do run the committee actively but depend on NGO/WASMO for many things
Gujarat (Bhuj)	70–50%	Women do play active role, however key decisions are made by the men in the committees
Maharashtra	70%	Women do not play any role but the committees follow government membership
Tamil Nadu I	30%	Very nominal role
Tamil Nadu II	30%	Very nominal role
MOUNTAINOUS		
Himachal Pradesh		No information: very nominal role
Meghalaya	0%	There are traditional women committees but no role or membership in VWSCs
Sikkim	30%	Traditional committees and women have their own committees and focus only on cultural and religious activities
Uttarkhand	30%	NGO makes an attempt to involve women in planning and implementation

In the case of northeastern state of Meghalaya, it is the traditional 'Dorbar' structure which takes care of the drinking water supply systems apart from their traditional local administration responsibilities. The Dorbar is equivalent to a Panchayat and is the recognised local body under the Scheduled Tribes Act. Women are not part of this Dorbar and they have separate committees which perform cultural activities and celebrate community festivals etc. However women are part of the committees constituted as per the government guidelines for different schemes. In Mawklot-Swajaldhara VWSC case and in Raitsalia Dong Water User Committee (according to the Soil and Water Conservation Department guidelines) women are members. While the Mawklot VWSC is inactive, the Raitsalia Water User Committee is active. In Raitsalia, women are responsible for maintaining the accounts of collecting the user charges and paying for electricity bills which they succeed in doing without records.

It can be concluded that the representation in many of the RWSS/PHED managed schemes is nominal while the NGO promoted/World Bank assisted projects do have greater women's participation. Membership in PHED/RWSS managed cases are confined to the government norms and women were confined to passive membership with their attendance at meetings and training best described as nominal. In NGO/World Bank assisted projects intentional efforts have been made to bring more women into the committees and office bearers as seen in Gujarat, Odisha and West Bengal. In these models NGOs, as part of the social mobilisation, created awareness and made sure that women were part of the committees.

Women's participation in planning, designing and implementation

Though women bear the drudgery of waiting, filling and carrying head loads of pots or buckets to meet their family's basic needs, they are hardly consulted during the design of the scheme. The myth that only men can understand the technical details seems to dominate in most of the RWSS/PHED led schemes while the NGO and World Bank promoted schemes did make an intentional effort to make women participate in the process. Gram Vikas developed gender-sensitive designs, requiring community agreement and delivery of three taps per household: one in the toilet, one in bathrooms and one in the kitchen – very much designed to meet the needs of women at the household level.

WASMO has taken extra efforts in creating an enabling environment during the planning stage of the project by involving women in focus group discussions and in the preparation of Village Action Plans. In Odisha (Gram Vikas), Punjab, Kerala and West Bengal similar efforts were also undertaken. In the case of Chatthisgarh, Madhya Pradesh, Maharashtra, Tamil Nadu and the Telangana cases (Naandi and others) efforts to involve were minimal.

In the case of Kerala, women in general have an overall understanding of the issues in water and have a strong voice to raise their concerns or give opinions. The 'People's Participation' in Kerala actually originated with women's participation in planning and development activities, through the 'ayalkoottams' (neighbourhood committees). This participation is now carried over to issues related to drinking water supply. Within the NSJVS (community service provider), every consumer is a member. When they become members all are made aware about their roles and responsibilities. Though the reservation system entails only 50 per cent of women in elected bodies and in other areas, in Kerala the women in such positions constitute more than 60 per cent.

In Meghalaya the woman's role is not visible. Traditionally, women are kept away from participating in any administration/implementation of development programmes for the reason that the men see it as their domain. According to the men, they take the views of women in all aspects but they do not want the women to participate in meetings or common forums along with men where the issues are discussed. The discussions revealed that the women are willingly staying away from such participation in common forums/meetings but if necessary they are ready to participate and they are capable of making decisions and implementing them as well. The women's view is that they can influence the decisions even if they don't participate in the meetings/discussions along with men.

In Sikkim, women are participating in discussions in the planning process mainly through the Ward Committees as part of decentralised governance. Discussions with them revealed that they are aware of the current issues and can actively contribute to the Ward Committees' activities.

Women's participation in decision-making processes

Across the 20 case studies it can be seen that women's role tended to be confined to membership and their participation in decision-making is limited. Women hardly attend meetings and some of them simply sign the registers without even attending the meetings. Most of the times the husbands act as phantom leaders and attend the meetings on behalf of their wives, leaving them with no opportunities to learn or obtain exposure to or awareness of the issues. Women also do not demand these opportunities as they are required to be busy with their household chores and agricultural operations. However there are exceptional cases such as Gujarat, Odisha (Gram Vikas), Karnataka and Kerala where women do play an important role in administrative decision-making.

The special efforts made by the organisations in these states in appointing women as training coordinators and social mobilisers to reach the women has borne fruit and women are able to take decisions and even approach government officials when needed. In Kerala all the ward members, executive committee members and governing body members follow a consultative decision-making process, which could be attributed to the advanced decentralised Panchayat system and high literacy levels. However in the other case of Kerala, the woman *sarpanch* (chair) is very active but somehow the water committees do not have women representatives. It can be seen from these cases that it not easy to break the stereotype of men dominating the water domain, but if proper measures are taken then women can excel in their roles and perform successfully.

Women's participation in financial management

Financial management is one area that women are hardly involved in, but wherever they are given the responsibility they are able to perform well. In Odisha (Gram Vikas) the women act as cashiers and people tend to believe that they are better custodians for financial management. In Singhpura in Punjab a female accountant is meticulously managing the receipts, payments and records. Women members in both Gandhinagar as well as the Kutch district of Gujarat are very good at information sharing and accountability, book-keeping, maintaining records of operation and maintenance, etc. Every activity is systematic and is a strong case for replication and sustainability with everything recorded on paper. In Shinay, a woman was recruited for the position of computer operator. Though she doesn't have an educational background in the subject, her interest encouraged her to learn the subject. Now, all the records are computerised. In the Telangana RO case study, both the villages supported by NCWS had women as operators. These two operators are trained and are well capable of handling both technical and other operation and maintenance issues. These operators are also responsible for financial transactions with customers and also maintain the financial records.

However, in most cases financial transactions and management are dominated by men, and women do not get any space to discuss the transactions. Though women have vast experience in self-help group management, their skills are not valued enough and women are confined to passive attendance. It can be noted that mere reservations in committee membership may not bring the desired change unless there are intentional efforts to build their capacity towards empowerment.

Women's participation in training and capacity building

In general there is no specific focus on training women specifically apart from mobilisation and awareness-building. But in Jharkhand the *bahus* (daughters-in-law) of the village are trained as water committee secretaries (*jalsahyyas*) in record-keeping and water quality testing. This effort of the rural development department in Jharkhand has resulted in building awareness among women such that they are able to effectively maintain the records. The PHED has prepared detailed guidelines for scheme operation in the form a book that the VWSC can refer to. However, this book has so far only been provided to the service provider in Bero. The VWSC president, vice president and *jalsahyyas* also receive formal training for their roles from PHED. The *jalsahyyas* are trained regarding book-keeping and water quality testing.

WASMO in Gujarat has been a forerunner in enhancing the capacities of women by facilitating training programmes for women's empowerment; in the year 2012–2013, a total of 133 such programmes were organised across the state. Apart from this, WASMO also takes the water committees to visit the best performing villages. The exposure visits encourage members to perform better and there is evidence of cross-learning.

In World Bank assisted programmes in Karnataka, Kerala and Punjab, the water committee members are trained, however there is no specific training for women. In most of the cases, the training is limited to awareness raising and there are no efforts to boost the morale of the women to participate in the water supply management.

Women staff members in the enabling support environment

After analysing the successful community-managed systems it was found that having women staff members in the enabling support entities is one of the key criteria to mobilise women. Staffing pattern across RWSS/PHED departments reveal that in Chhattisgarh there is only one woman engineer in the whole department and information from staff from the 15 states who were invited for the research dissemination workshops revealed that there are only 0.5–1 per cent of female engineers working in RWSS/PHED departments.

Some of the reasons cited were that women do not like field-based positions (field trips cause disturbance to the household responsibilities conventionally associated with women across much of Indian society), and have a preference to stay near to the towns, in addition to the traditions and cultural norms and demotivation from family members etc.

Further, some of the women officers expressed that they prefer to work in administrative positions which do not require travel. They also feel that it can be challengeing for women to deal with politicians, local leaders and contractors on certain issues. However, we note that the times are changing, there are more and more young women engineers coming to state departments as part of a new recruitment drive. There is then an urgent need to train these female officials to bring about the desired change in the mobilisation strategy for mainstreaming women. Some of the information from the states shows attempts to create gender-sensitive training platforms, venues and themes which encourage women to participate actively.

Overall we have found that gender balance and equitable opportunities can be achieved only when there are intentional efforts being made by the enabling

Table 13.2 Women staff members in the enabling support environment

State	Total engineers	Women engineers	Percentage of women engineers
Manipur★	287	33	11
Mizoram★	85	10	12
Nagaland★	211	10	5
Punjab	879	9	1
Rajasthan★	2281	397	17
Uttarkhand	337	5	1
West Bengal	784	17	2

Note: asterisk indicates unverified.

support organisations. Based on the successful cases, one can conclude that there is a need for special efforts both in staffing patterns and strategies to promote women by providing special platforms to express their opinions and voice their preferences. Having reservations in memberships does help but more efforts are needed to convert the passive participation to active performance, enhancing women's capacities through training. Women proved to be better book-keepers, water tariff collectors and custodians of the assets hence these capacities of women have to be carefully nurtured in future programme preparation.

Conclusion

Women have long been associated with community management. Often they hold positions as trustworthy and valuable members of water committees. This research reinforces this claim by showing that women commonly play a critical role at the community-service-provider level. However, it also highlights challenges, particularly in terms of the representation and role of women at the enabling-support-environment level, which continues to be male dominated across many of the case studies.

14 Discussion and conclusions about community management in India and beyond

This final chapter discusses the findings from this research and considers their implications for community management in India and internationally. It is divided into three major sections. It starts by discussing the research finding from an international perspective. This is followed by a discussion about the particularities of the Indian sector where a new framework focused on the 'coproduction' of rural water services is discussed. The chapter ends with some final conclusions considering what the research means for community management in the context of the Sustainable Development Goals (SDGs) for water.

Community-managed rural water supply in India in an international perspective

Having seen the variety of models for community management and support across India, the costs and the ultimate outcomes in terms of service level, this chapter aims to put these findings into international perspective. It will thereby focus on the differences and similarities between the successful approaches to community management in India and elsewhere, and with that, the potential of the findings in this book to contribute to the global debate on community-managed rural water supplies.

The chapter starts by identifying the groups of countries that are in a similar situation with respect to rural water supply as India as a whole, and for the individual states. It will do so by analysing coverage data in relation to GDP. It then continues by comparing the institutional models found in this study to the ones typically found in the groups of countries identified previously, both for the external support entities and the community service provider arrangement. It then continues by comparing the ultimate outcome: service levels between the Indian states and comparator countries. It ends by comparing the costs of rural water supply found in this study to the ones found elsewhere, again with emphasis on the groups of countries identified previously.

We have made similar graphs for the other countries in the world, using data from the Joint Monitoring Programme (JMP) for Water Supply and Sanitation (WHO and UNICEF, 2016). Figure 14.1 shows rural water supply coverage levels of countries in relation to their per capita GDP (PPP) in 2014 US dollars (World

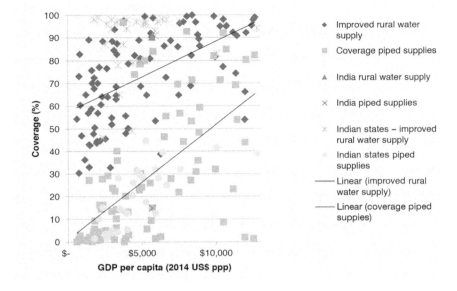

Figure 14.1 GDP per capita of India and Indian states

Bank, 2016), including only those countries with a GDP of less than $13,000 (as the state with the highest GDP per capita is Goa with around $12,700 per capita). It also puts the all India average and the individual states there. That shows that India's rural water supply coverage levels are above the international trend line. In other words, it has a relatively high rural water coverage for its level of GDP. Particularly notable is that many of the poorest states have very high levels of coverage for their level of GDP.

It also plots access to what the JMP calls 'piped on premises' (i.e. piped supplies with household or yard connections), in relation to per capita GDP, both of Indian states[1] and other countries in the world. Figure 14.1 shows that for piped on premises, India is a bit below the international trend line. So, even though India has a reasonable level of coverage with piped supplies, much of that still is through standposts, and not through household connections, at least as compared to countries with similar levels of wealth. Its current drive to move towards piped supplies and household connections is thus therefore well in line with what may be expected for its level of wealth. Particularly in the poorer states this will need to speed up significantly, as those are most below the international trend line.

It is also useful to look at who its nearest comparators are, both in terms of similar levels of GDP per capita and similar levels of rural water supply coverage. Comparing the all India GDP data, as well as the data of those states that we classified as middle income, provides a list of the poorer countries in Central America (like Honduras and Nicaragua), Asia (Pakistan and Viet Nam), eastern Europe and central Asia (Moldova and Uzbekistan), and the richer countries of Africa (such as Ghana).

The group we identified as neo-patrimonial or low-income states would be best compared to other Asian countries like Bangladesh, Cambodia and Nepal, which have similar levels of GDP, and still high levels of rural water supply. Many other countries that have similar levels of GDP have rural water supply coverage levels that are well below these Indian states. The GDP per capita of many African countries – that are sometimes compared to these low-income states – is well below that of the poorer Indian states. Only the two poorest states in India – that we didn't look at in this study – Bihar and Uttar Pradesh, have GDP levels as low as many African countries – but rural water supply coverage levels that are well above the ones typically found in Africa. The high-income states would have as nearest comparators those in Latin America (Bolivia and Guatemala) or Asia (Bhutan and Philippines).

The comparison of GDP, rural water supply coverage and coverage with piped supplies of Indian states with other countries in the world leads to two principal conclusions that are of relevance for the remainder of this chapter.

First, India has been able to achieve almost universal rural water supply coverage, and in that is doing better than countries of similar levels of wealth. This is particularly the case for the poorest states. In spite of all the challenges and limitations mentioned in previous chapters of this book, the models for rural water supply employed in India have been relatively successful in extending coverage in international comparison. Second, for other rural water supply performance indicators than coverage alone, India is best compared to its neighbouring countries in south, central and southeast Asia and ones with similar levels of wealth in for example Latin America.

The following section explores how the institutional model of external support entities with community service providers in India is similar to and differs from the arrangements elsewhere, particularly in comparable countries.

External support entities and their reform

Lockwood and Smits (2011) in their review of institutional arrangements and reforms for rural water supply in 11 countries from across the world, classify first of all, the degree and speed of decentralisation of both the service authority function and the corresponding financial resources. In that classification, India was considered to fall mainly in two categories:

1 partial decentralisation, whereby some functions are decentralised to local government, whereas others are retained by a central agency; and
2 inadequately resourced decentralisation categories, whereby functions are decentralised, but not the corresponding funding.

The findings in this study have confirmed that most cases are ones of partial decentralisation. The SRWSA-type agencies, even the reformed ones, retain significant powers during the capital intensive implementation phase. These agencies lead and finance large-scale implementation programmes and take the main decisions on the way in which these programmes are executed. Local governments have a very small role in capital investments. In some states (such as Kerala or Tamil Nadu), however, the service delivery role has been decentralised to Gram Panchayats, and Gram Panchayats contribute financial resources to service delivery.

This partial decentralisation is common in other comparator countries such as Ghana or Honduras. In those countries, central agencies still lead and finance most of the implementation works. WaterAid (2008) found in a study in 12 countries that two-thirds of budgets for water remained outside local government control, but with central agencies.

In addition to decentralisation, Lockwood and Smits (2011) argue that it is also important to look at other dimensions of the reforms:

- Separation of the authority and provider functions. This means that local governments have received the responsibility for the authority functions of planning, monitoring, oversight and on-going support. Providers are the entities responsible for day-to-day operation, maintenance and administration, and can either be community-based organisations, small-scale private providers or public ones. Such separation of functions is clearly defined in countries such as Colombia, Honduras or South Africa, and considered key to ensuring accountability. Most cases we found in India differ significantly in that sense, because the VWSCs, as the most commonly occurring type of community service provider, are a sub-committee of the Gram Panchayat, the local government. In fact, the national guidelines for drinking water provision prescribe that the president of the Gram Panchayat also acts as the chair of the VWSC. And we see that the operations and administration of the VWSC and Gram Panchayat run through each other. For example, Gram Panchayats provide financial and human resources to VWSC. This may give the VWSC more strength and capacity, but it also may compromise effective oversight: Gram Panchayats become judge and jury.

- Regulatory authority. Another reform observed is that several countries have set up independent regulatory authorities for water supply, initially focused mainly on urban areas, but increasingly trying to do 'light-touch' regulation of rural (community-based) service providers. This is the case in Latin American countries, such as Colombia and Honduras. In the review of cases in India, we haven't come across such cases of having independent regulatory authorities for rural water supply.

- Process of reform of central agencies. Next to the result of the reform, it is important to pay attention to the process of reform. As functions for rural water supply decentralise, central agencies either resist or co-opt those reforms. The clearest manifestation of that is their retaining of the implementation role (i.e. leaving decision-making for service provision to local government, but keeping the technical – and financially important – role of infrastructure implementation). In that way, the experiences we reviewed in India resemble experiences with reform of central agencies, as in Ghana or Sri Lanka. In some countries, central agencies have actively contributed to the reform processes by taking on new tasks, particularly around post-construction support, as in Honduras. This we have only seen in some of the cases in India, such as Gujarat.

Community service providers

Where India differs more from similar countries is in the service provider arrangements. Many of the cases presented in this book can be classified as one of 'direct public provision with community involvement'. PHED-type agencies fulfil many of the service provision roles, and VWSC have only minimal executive roles, acting as liaison with the PHED or doing tariff collection. This modality of community management is largely absent in other countries, for example those reviewed by Lockwood and Smits (2011). They only found public provision in small towns, and through more formal municipal utilities.

In most other countries reviewed by Lockwood and Smits (ibid.), the common modality is one of the 'community management with support', and particularly in the wealthier countries venturing in more professionalised community service providers, such as Colombia or Sri Lanka. In India, we found the modality of 'community management plus' mainly in NGO-supported programmes and in the more remote mountain states, whereas elsewhere it is often the main model. The more professionalised community-service providers were only found in the wealthier states, thereby coinciding with the trend elsewhere.

All in all, the institutional set-up for external support entities to rural water supply in India coincides in some important ways with similar countries. In particularly, partial decentralisation whereby a central agency retains the implementation role but local government is responsible for support to service delivery, is commonly found in other comparable countries. Where India differs markedly from other countries is in the unclear separation of the authority and provider functions. Gram Panchayats, as local authorities, oversee the VWSC, but also have an executive role in service provision.

India also differs much from the service provider arrangements elsewhere. The modality of direct public provision with community involvement is not common elsewhere, whereas in India the modality of community management with support is not so common. Where India and its comparator countries do coincide is in the trend that as states become wealthier, there is a shift towards more professionalised service provision.

Conclusions about the relevance of the Indian experience for the international context

India has achieved very high levels of rural water supply coverage, currently at 96 per cent. With the exception of the mountain states, all states have similarly high levels of coverage of more than 90 per cent. When it comes to piped supplies, however, coverage tends to correlate with the per capita GDP of the state – with barely any coverage through such supplies in the poorer states.

Comparing these data internationally, India is ahead of the curve as far as coverage is concerned, but slightly below the curve with respect to piped supplies. In other words, for its level of per capita GDP it has a high overall coverage, but a relatively low coverage with piped supplies. The analysis has helped identify countries that have a similar situation in terms of rural water coverage and coverage with piped supplies in relation to the GDP. This puts India – and its different groups of states – among other Asian countries in the neighbourhood (such as Bangladesh, Bhutan, Cambodia and Pakistan), Latin American countries (Bolivia, Guatemala, Honduras and Nicaragua) and ones in eastern Europe and central Asia (Moldova and Uzbekistan). The closest comparator in Africa is Ghana.

Comparing the institutional models found in India with those found in some of the comparator countries shows some similarities. Also several of the comparator countries have partially decentralised rural water supply, in the sense that service delivery responsibilities have been transferred to local government, but the role for infrastructure development has remained with centralised agencies. Local governments still have in many of these countries very limited funding for investments and central agencies play an important technical role. Where India differs from the comparator countries is in the models for community management. In several of the states studied in this research, direct public provision with some community involvement was the most common model – whereas elsewhere the classical 'community management plus' model is more common. Also, in India there is no clear separation of the service authority and service provider roles, one that is more pronounced elsewhere.

It is difficult to compare the ultimate result of these service delivery models and service levels across countries. In our study, we specifically focused on the most successful cases, and the ways in which service levels are defined differ across countries. In spite of these caveats, the service levels found in the 20 cases are indeed good in international comparison, with about 30 per cent of the interviewees failing on only one of the service-level indicators – whereas often it is only 20–30 per cent of the population that gets an adequate level of service.

The costs of providing these levels of service are even more difficult to compare. The spread of unit costs for capital investments both across the 20 Indian case studies and across comparator countries is very large. But by any measure the capital costs in India in this study were higher than previously reported on suggesting India is not as low-cost as previously understood. Recurrent costs on the other hand fall in the few ranges for reference costs that exist. More important than the exact cost level is the fact that around 50 per cent of these costs are not paid for by users, but by external support entities. This shows the important contribution of public finance to covering these recurrent costs. These have, among others, enabled the mentioned service levels.

By putting these findings into international perspective, a number of implications for the global debate on community-managed rural water supply can be formulated:

- Communities play a key role in managing their rural water supplies, but cannot do it on their own. This was the hypothesis when we set out the research, and this is largely confirmed. Both centralised PHED-type agencies and local government were found to play an important support role to community service providers.
- Institutional arrangements for rural water supply therefore need to clarify where those support roles lay, both during the capital intensive infrastructure development, and during the actual service delivery phase. This is particularly of importance where rural water sectors go through reform processes of their central agencies.
- An important implication from the international debate for India is the importance to better separate the authority from the service provider functions, for accountability reasons. Local governments elsewhere tend to have more of an oversight role. This allows local governments to be involved, but keeps a good distance to play that oversight.
- Support to community-managed rural water supply is not limited to 'soft' activities of monitoring or technical assistance, but can also come in the form of funding parts of the recurrent costs. In many comparator countries, there is supposed to be 'full cost recovery', but in reality many of the capital maintenance and direct support costs are paid for by external support entities and not out of tariffs. The Indian examples, on the other hand, show how the external entities actually do cover these costs. A first step in defining who could or should pay for these costs is making a full and complete assessment of what these costs are, as was done in this study.

Explaining community management as a form of coproduction

While there are some similarities between India and the international context, the Indian context also has some idiosyncrasies that are specific to that context, in particular the overlap between the Gram Panchayat and community management. This section therefore discusses this overlap in more detail by suggesting it may be better to describe this approach as a form of 'coproduced service delivery' between state and communities and that this type of shared approach may offer an effective model for reforming community management in other contexts. This argument starts from the position that although the called for reforms towards community management plus are considered radical in terms of recasting responsibility for service delivery as a shared responsibility between governments and communities (Baumann, 2006) they are, discursively, reformist in scope. That is, they have retained the discourse of 'community management' and continue to be limited by the traditional distinctions of the public-private divide. Yet in India the dominant forms of service delivery documented throughout this study exhibit institutional and financial costing arrangements that reflect a balance of recurrent inputs from the state (or other agencies) and communities that challenge that public–private divide and, more broadly, the discourse of community management.

In explaining the arrangement it is tempting to still be limited by a binary discourse with a continued emphasis on either describing it as a form of community management when the community takes the lead, or in cases where Gram Panchayats are more prominent, then a form of direct public provision (Rout, 2014). Yet the findings from across the case studies consistently suggested a more nuanced situation with water committees working within, alongside and with approval from local self-government in nearly all 'community management' programmes captured in this study. The services provided are also recurrently financed between a balance of tariffs and public subsidy (taxes) and so, in response, this section helps to conclude the book by introducing the concept of institutionalised coproduction (Joshi and Moore, 2004) as a way to conceptualise rural water services in India and – potentially – broader contexts.

The introduction of this concept forms a modest attempt to suggest a shift in discourse beyond community management, which promotes a sense of equivocation or, even, contradiction when used by governments, NGOs and donors to describe the programmes that *they* deliver. Even in the conventional community management paradigm, external agencies played a critical role in leading the implementation of schemes, including finance, construction and capacity building, but with ambiguous on-going support arrangements. Community management plus was an attempt to emphasis the on-going responsibility of the governments, NGOs and donors to those programmes in terms of recurrent monitoring, finance, technical support and capacity building (Lockwood and Smits, 2011). Yet this is considered to reflect a sort of discourse-lag, in which the conventional discourse of the sector has been reformed rather than changed. This suggests that the status quo is acceptable rather than highlight the importance of fully reconceptualising how rural water services should be delivered. In this sense, the coproduction discourse, which through its basic definition indicates a shared role for governments and communities, is considered such a potentially useful route for shifting sector thinking beyond the 'limits' of community management as an approach.

The basis of the coproduction concept – which has been applied in both the developed and developing world (Bovaird, 2007; Bovaird et al., 2015; Cepiku and Giordano, 2014; Joshi and Moore, 2004) – is to describe a form of public service provision that involves the sharing of resource, labour and responsibility between public agencies and citizens. However, beyond that basic notion, Joshi and Moore (2004, p. 50) develop four particular characteristics to characterise it more precisely within the context of low- and middle-income countries, in order to develop their specific sub-concept of institutionalised coproduction:

1   service delivery that involves substantial resource contribution from the state and private citizens;
2   service delivery that is based on long-term relationships between the involved parties;
3   service delivery that has the potential for informal arrangements governing those relationships; and

4   service delivery that involves a blurring of distinctions between the traditional divide of public and private actors.

This section now considers each of these four points in turn. The discussion presented is intentionally framed as an examination of the dominant trends across the case studies and thus seeks to explain the most common arrangements. This is particularly in relation to the 17 case studies that involved water committees that were either part of the Gram Panchayat or were registered under them.

The research investigated the financial costs of services across the case studies, which provides evidence regarding the balance of resource contribution from between the state and citizens. It was explained in Chapter 2 that the most common 'default' cost-sharing arrangement for community management involved communities covering 100 per cent of the on-going OpEx for rural water services (Joshi, 2003). Taking direct OpEx across all the case studies, the findings reported in Chapter 11 show that communities cover 63–94 per cent (IQR) of OpEx costs with support agencies providing subsidy for the remaining 6–37 per cent (IQR). This means that significant subsidies cover everyday costs, such as energy and bulk water. Beyond direct operational costs, there are OpEx Enabling Support costs which are covered by support agencies so, overall, 7–48 per cent (IQR) of the every-day-recurrent costs of services come from external agencies. The evidence that is provided therefore clearly indicates that both citizens and support agencies – usually government – provide substantial recurrent resource contribution in these success-ful cases of 'community management' in India. Due to the extent and consistency of resource sharing, these arrangements are considered to be more appropriately framed within the institutionalised coproduction concept (Joshi and Moore, 2004).

The longevity of the relationship between the state and citizens is another char-acteristic highlighted by Joshi and Moore (ibid.). At a basic level, the average length of operation across the cases was seven years which helps to verify that the service delivery arrangements studied are durable at least over the medium term. Yet more broadly, part of the challenge of shifting from community management to commu-nity management plus, is related to developing a bipartite sense of responsibility between communities and support agencies over the long term (Lockwood, 2002, 2004). What this research indicates is that there is a high degree of permanence in institutional relationships within government programmes in India as they have been embedded and partially prescribed within constitutional reforms from the early 1990s (Banerjee, 2013; Government of India, 1993). That constitutional amendment prescribes that the local self-government of the village – the Gram Panchayat – is ultimately responsible for the local management of rural water serv-ices (Government of India, 1993) and successive policy programmes have sought to maintain this principle while promoting community management (Government of India, 2003, 2013a).[2]

This means that the water committee can be an official sub-committee of the local self-government, as shown in nine case studies. Within the government system, it is also possible to have a water committee that is a separate registered society (as found in eight case studies) yet these only exist with official approval and oversight from the Gram Panchayat. It was only in three case studies documented in this research, that the Gram Panchayat did not have an official role within service provision, with this described as the NGO-influenced unregistered society model. Despite this last group, the dominant trait across the case studies was for the institutionalisation of 'community management' within the government system. This is considered to compare favourably to the model of community management that had become common in some parts of the world whereby external agencies such as NGOs have a minimal relationship with communities beyond the implementation phase of projects, or even no relationship at all (Van den Broek and Brown, 2015; Harvey and Reed, 2006). The Indian experience suggests that for developing permanence in relationships between citizens and support agencies it is most likely to be beneficial to root such arrangements within the (local) government system. Conceptually, however, this shifts the arrangement from one which is best described through the prism of community management to a set-up more accurately described as a form of coproduction.

Part of the arguments developed by Joshi and Moore (2004) is that despite having some level of permanence the arrangements for institutionalised coproduction can be based on informal relationships. This research suggests that this is rare in India with it only being found in the three unregistered society case studies described previously, with the remaining 85 per cent of the cases being recognised formally within the legislative framework of Indian law. In this sense, the informality proposed by Joshi and Moore (ibid., p. 50) is not recognised in this research at this legalistic level, however they position this as a tendency rather than a rule, specifying that 'institutionalised need not involve the kinds of contractual or quasi-contractual arrangements between state agencies and organised non-state actors'. Therefore, this is not considered to invalidate the applicability of the institutionalised coproduction concept in describing the general patterns of institutionalised arrangements found across the case studies.

The final key criterion associated with institutionalised coproduction is that it 'implies blurring and fuzziness in the lines that Max Weber, in particular, taught us to try to define clearly and precisely: the boundaries between public and private' (ibid., p. 50). This is considered to be especially applicable to describing the public-private institutional set-up found in the case studies where the VWSC is an official sub-committee of the local self-government. Under this arrangement as prescribed in the official rural water supply policy (MDWS, 2013), the water committee should have between nine and twelve members. However, at least two of the most important members of the water committee are employees of the local self-government and their role is considered part of their duty as a public servant. These two members include the President of the Gram Panchayat who operates as the water committee chairman and the Secretary of the Gram Panchayat who is the treasurer of the committee. In this sense, much of the key labour contribution comes through professionalised public servants with the remainder of the committee made up of private citizens. The level of contribution from the private committee members can be limited to attendance of committee meetings, rather than direct labour contributions towards the operation, maintenance or administration of the system. Beyond the chairmanship and book-keeping, undertaken by the public servants, the most labour-intensive role is the 'pump operator' which is undertaken by a private citizen directly appointed by the committee. This individual is paid a wage agreed by the committee but together with the public servants it means that the individuals contributing towards service delivery tasks are all remunerated for that purpose. This again helps distinguish the Indian experience captured in this research with the implicit volunteerism that is associated with the community management model in an international context (Moriarty et al., 2013). In the context of the institutionalised coproduction concept, the arrangement shows how services are delivered through an institutional structure that combines public servants, private employees and voluntary members leading to a model that blurs the line between public provision and private community management.

Community management and the Sustainable Development Goals

This chapter has highlighted the international implications of the research and suggested that the dominant institutional models found in the research can be described as a form of institutionalised coproduction. This section now considers the research's implications for the policies and practices associated with rural water services. For this purpose, this chapter starts by framing the discussion in the context of SDG Goal 6 to 'ensure availability and sustainable management of water and sanitation for all' (United Nations, 2015). That goal commits the global community to a number of targets and principles of which two are particularly relevant in terms of this research:

- *Target 6.1*: By 2030, achieve universal and equitable access to safe and affordable drinking water for all;
- *Target 6.b*: Support and strengthen the participation of local communities in improving water and sanitation management.

Target 6.1 has some particularly critical words in terms of shaping the ambitions of the global water community. They are considered to be *universal, equitable, safe* and *affordable*. This research has findings that are especially relevant for the debates about safe and affordable drinking water while it also provides findings that have secondary implications for debates about universal and equitable services. It also has implications for Target 6.b in terms of the participation of local communities which will be addressed later.

Initially, focusing on the notion of safe water services there is evidence that many improved water supply services do not provide technically safe water services (Clasen, 2012; Onda et al., 2012; Parker et al., 2010). This research did not attempt to measure water quality but through the service level approach provided a more sophisticated measure of different components of water services that contribute to their safety, as compared to the basic measure of improved and unimproved access used in the MDGs (Burr and Fonseca, 2013). As reported in Chapter 11, the analysis showed that people with household connections were significantly more likely to have higher service levels, than any other water access point – over 70 per cent of respondents with this type of access in programme villages reach at least basic on the composition service-level indicator. In comparison, when taking data from the one case study with handpumps, 0 per cent of respondents reported receiving a service that could be classified as basic on the service-level ladder.

The research can therefore be considered to show that rural water service policy around the world should have the ambition of delivering piped water supply with household connections in order to deliver high service levels. This reinforces the timeliness of the Indian policy shift in 2013 (Government of India, 2013a) to concentrate resources on piped water supply with household connections. Yet in terms of a wider context, Sub–Saharan Africa remains the region with the biggest overall need in terms of rural water services (WHO and UNICEF, 2013). In that context community management had become associated with handpumps, which are the most common form of water system in rural areas (Van den Broek and Brown, 2015). The next generation challenge in that continent will be to move people from handpumps to piped water supply and so a key policy-related finding from India is that the forms of service delivery studied here can play a role in this transition.

However, with increased technical sophistication, comes increased cost. Through the cross–case study analysis methodology this research was able to reveal many 'hidden costs' related to rural water services, such as the subsidised power costs that rural water service providers receive in several states. It also revealed a diverse range of different funding mechanisms that feed into rural water services coming from various layers of government and specific funding streams. It is contended that previous research underestimated such costs by relying largely on survey data for collecting primary data on costs (Burr and Fonseca, 2013), rather than the key informant interviews that feed into the case studies for this work. In this sense, the research has demonstrated that the financial costs of delivering high-quality rural water services in India are higher than the widely used sector benchmarks indicated (ibid.; McIntyre et al., 2014).

In the context of the SDG target 6.1, recognising the higher than previously recognised costs associated with successful services means that the affordability of services to all in society is a significant challenge. In India the solution to this problem has been for government to leverage the capability of the local self-government and to subsidise services either directly or indirectly. This is considered a critical lesson from this research: rural water services require significant levels of subsidy if they are to be successful – a finding that has been recently recognised in other studies (Franceys et al., 2016). Mobilising sufficient funds for public investment is considered to be critical for ensuring the related challenges of equitable and universal services are achieved. Yet with public fund mobilisation limited in most low- and lower-income countries (Norman et al., 2015), ensuring users that are able continue to pay for at least some of the costs through user charges is still likely to be important going forward. This balance of funding, as well as other forms of joint contribution between the state and communities, is considered critical for the future of the sector and why the notion of institutionalised coproduction was introduced earlier in this chapter.

In that spirit, moving onto SDG target 6B on the need to support and strengthen community participation in water (and sanitation) management, this research raises a sceptical note. This is not because of a belief that community participation is bad for rural water services but rather that community participation is not a necessary condition for success. This observation is made in two regards. First, the data from this research shows no definite relationship between the observed community participation level and service level outcomes. Second, the emphasis – almost fetishism – of participation in rural water supply is considered to be misplaced in terms of achieving universal coverage as per the SDG target 6.1. There are many communities – or parts of communities – where active participation in rural water services is problematic due to issues such as social structures of patronage and exclusion (Nelson and Agrawal, 2008) or, simply, some community members having limited capacity to be effective managers of water services. Based on the evidence presented and the explained logic of reasoning, the researchers support a change in emphasis in policy and practice around community participation. Communities should still be involved but there are different ways for this to happen – either through highly devolved democratic systems, conventional participatory approaches or shifting towards an urban-consumer type arrangement. Key to the shift though is that whatever role communities take in particular programmes there needs to be an emphasis on greater responsibility sharing with governments and other external agencies is considered the most critical implication of the research.

Offering a final conclusion, the research started by asking what does successful community management look like in contemporary India? And how much does it cost to support it? The research showed that there are different community management models with various community service provider and enabling support environment arrangements but that a common pattern across the cases was significant level of continued support from local self-government and other support agencies to VWSCs. This was through both direct technical and administrative support services but also significant recurrent financing of service provision costs. In many instances, the level of responsibility sharing between government and communities is considered to reflect a shift in the practices of community management as India has undergone significant economic transition towards a model characterised by the coproduction of rural water services between state and citizen. This route of transformation has largely be driven by domestic policy contexts but is still considered to provide an insight for other countries undergoing significant economic transition about how community management can be more effectively professionalised to provide better and more reliable water services. It requires states, NGOs and other agencies to ensure they do not just leave communities isolated but instead institutionalise a system of support that can provide recurrent support – administrative, technical and financial – to rural communities over the long-term.

Notes

1   The Indian census only identifies people with piped supply, not 'piped on premises'. Piped supplies with household connections constitutes about half of all forms of piped supplies across India. We have therefore assumed also for individual states that half of their piped supplies is with piped on premises.
2   The reconciliation of the domestic devolution agenda with what could be described as the more internationally influenced ideals of community management is considered critical to understanding the Indian context and is a point further examined later.

Appendix

Overview of case studies and analytical tables

Table A.1 State-wise rural water supply and development indicators dataset

State	Improved water (rural)	Household piped water supply (rural)	GDP per capita (PPP)	Human Development Index	Devolution Index (rank)	Below poverty line (rural)	Literacy rate	Gini	Growth in poverty elasticity	Physio-graphic zones
All India	96%	31%	$4,243	0.467	n.d	26%	74%	32.3	-1.2	Other
Andhra Pradesh	97%	63%	$4,643	0.473	13	11%	92%	32.9	-6.1	Other
Arunachal Pradesh	80%	59%	$4,876	0.573	20	39%	67%	32	-1.7	Mountain
Assam	87%	7%	$2,525	0.444	16	34%	73%	23.8	-2.2	Mountain
Bihar	98%	3%	$1,780	0.367	17	34%	64%	22	-1.9	Other
Chhattisgarh	97%	9%	$3,340	0.358	9	45%	71%	27.5	-2.5	Other
Goa	94%	78%	$12,786	0.617	16	7%	87%	27.6	-1.2	Other
Gujarat	97%	56%	$6,094	0.527	10	22%	79%	30.1	-1.2	Other
Haryana	97%	64%	$7,611	0.552	11	12%	77%	25.3	-1.9	Other
Himachal Pradesh	96%	89%	$5,265	0.652	12	8%	84%	27.4	-2.7	Mountain
Jammu & Kashmir	78%	56%	$3,382	0.529	n.d	12%	69%	23.9	-1.2	Mountain
Jharkhand	96%	4%	$2,632	0.376	21	41%	68%	27.4	-2.1	Other
Karnataka	96%	56%	$5,108	0.519	2	25%	76%	30.8	-1.3	Other
Kerala	93%	25%	$5,922	0.79	1	9%	94%	30.1	-2.8	Other
Madhya Pradesh	98%	10%	$2,955	0.375	4	36%	71%	27.4	-2.2	Other
Maharashtra	98%	50%	$6,679	0.572	3	24%	83%	34.8	-1.7	Other
Manipur	46%	30%	$2,372	0.573	18	39%	80%	16	-2.9	Other
Meghalaya	65%	29%	$3,511	0.573	n.d	13%	75%	21.6	-3.7	Mountain
Mizoram	49%	41%	$4,342	0.573	n.d	35%	92%	23	-1.2	Mountain
Nagaland	79%	52%	$4,423	0.573	n.d	20%	80%	19.1	-2.1	Mountain
Odisha	94%	8%	$2,998	0.362	13	36%	73%	30.7	-3.4	Mountain
Punjab	97%	35%	$5,268	0.605	19	8%	77%	27.2	-7.5	Other
Rajasthan	87%	27%	$3,763	0.434	6	16%	67%	26.8	-1.0	Other
Sikkim	83%	83%	$10,068	0.573	8	10%	82%	24.8	-1.1	Other
Tamil Nadu	98%	79%	$6,427	0.57	5	16%	80%	33.1	-1.6	Mountain
Tripura	94%	25%	$3,976	0.573	10	17%	88%	32.6	-2.2	Other
Uttar Pradesh	99%	20%	$2,068	0.38	15	30%	70%	28.1	-1.9	Mountain
Uttarakhand	91%	64%	$5,916	0.49	14	12%	80%	29.8	-2.5	Other
West Bengal	98%	11%	$3,997	0.492	7	23%	77%	32.4	-4.3	Mountain

Table A.2 Ladder of participation in key issues in rural water supplies

Type of community involvement	Phase in service delivery cycle			
	Capital investment phase	Service delivery phase	Capital maintenance phase	Service enhancement or expansion phase
1. Self-mobilisation	The community practises self-supply and seeks to improve this, or have developed an implementation plan and seek external support.	The community take responsibility for administration, management and operation and maintenance, either directly or by outsourcing these functions to external entities.	The community practises self-supply and invests in asset renewal, or identifies need and seeks external support for asset renewal.	The community practises self-supply and invests in service enhancement or expansion, or identifies need and seeks external support for service enhancement or expansion.
2. Interaction participation	The community in partnership with the service provider and/or support entities engage in a joint analysis of implementation options before developing a plan.	The community in partnership with the service provider and/or support entities engage in joint decision-making regarding appropriate arrangements for administration, management and operation and maintenance.	The community in partnership with the service provider and/or support entities engage in joint decision-making regarding asset renewal.	The community in partnership with the service provider and/or support entities engage in joint decision-making regarding service enhancement or expansion.
3. Functional participation	The community is provided with a detailed implementation plan that they discuss and they have a chance to amend limited elements.	The community is provided with administration, management and operation and maintenance arrangements that they discuss and they have a chance to amend limited elements.	The community is provided with an asset renewal plan that they discuss and they have a chance to amend limited elements.	The community is provided with a service enhancement or expansion plan that they discuss and they have a chance to amend limited elements.

Table A.2 continued

Type of community involvement	Phase in service delivery cycle			
	Capital investment phase	Service delivery phase	Capital maintenance phase	Service enhancement or expansion phase
4. Participation by consultation	Community members are asked whether they want a predefined implementation scheme but have no formal decision-making power to demand alternatives.	The community discusses administration, management and operation and maintenance functions but have no formal decision-making power to demand alternatives.	Community members are asked about asset renewal but have no formal decision-making power to demand alternatives.	Community members are asked about service enhancement or expansion but have no formal decision-making power to demand alternatives.
5. Passive participation	Community members are informed that project implementation is going ahead as per an externally designed plan.	Community members are informed how administration, management and operation and maintenance will operate without opportunity for changes.	Community service provider informs community members about asset renewal as per an externally designed plan.	Community service provider informs community members about service enhancement or expansion as per an externally designed plan.

Sources: Pretty (1994); Adnan (1992); Smits et al. (2015).

Table A.3 Organisational partnering typology for relation between enabling support environment and community service provider during different phases of service delivery cycle

Type of partnering	Phase in service delivery cycle			
	Capital investment phase	Service delivery phase	Capital maintenance phase	Service enhancement or expansion phase
Collaborative	ESE and CSP share responsibility for decisions regarding hardware (e.g. infrastructure) and software (e.g. capacity building) development during implementation.	ESE and CSP share responsibility for decisions regarding administration, management and operation and maintenance.	ESE and CSP share responsibility for decision-making regarding asset renewal.	ESE and CSP share responsibility for decisions regarding service enhancement or expansion.
Contributory	ESE and CSP pool financial resources to meet the costs of capital investment in hardware and software provision during implementation.	ESE and CSP pool financial resources to cover costs of administration, management, and operation and maintenance.	ESE and CSP save and pool financial resources to meet the costs of asset renewal.	ESE and CSP save and pool financial resources to meet the costs of service enhancement or expansion.
Operational	ESE and CSP work together contributing labour and/or resources to deliver hardware and software provision during implementation.	ESE and CSP work together contributing labour and/or resources to support administration, management, operation and maintenance.	ESE and service provider contribute labour and/or resources for asset renewal.	ESE and CSP contribute labour and/or resources for service enhancement or expansion.
Consultative	ESE and CSP communicate regularly during implementation with structured opportunities for feedback and dialogue.	The ESE and CSP have a systematic and transparent system for sharing information regarding administration, management, and operation and maintenance.	ESE and CSP systematically share information regarding service levels and technology status enabling proper planning for asset renewal.	Information regarding service levels, technology status and population is systematically shared, enabling proper planning for service enhancement or expansion.

Table A.3 continued

Type of partnering	Phase in service delivery cycle			
	Capital investment phase	*Service delivery phase*	*Capital maintenance phase*	*Service enhancement or expansion phase*
Transactional	ESE and CSP initially negotiate an implementation plan that is then delivered by the ESE.	The ESE and CSP fulfil different elements of the administration, management, and operation and maintenance functions as per negotiated arrangements.	Asset renewal is dependent on negotiations between ESE and CSP following a request from the CSP.	Service enhancement or expansion is dependent on negotiations between ESE and CSP following a request from the CSP.
Bureaucratic	ESE provides CSP with a standardised model of hardware and software provision during implementation.	Bureaucratic standards dictate the system for administration, management, and operation and maintenance.	Asset renewal is dependent on generic programme timelines (i.e. every *x* years).	Planned asset replacement, expansion or renewal is dependent on generic programme timelines (e.g. every *x* years and/or with every *x*% of population increase).

Sources: Demirjian (2002), Smits et al. (2015).

Table A.4 Summary of findings on organisational arrangements across the case studies

| Case | State | Organisational typologies | | Professionalisation | | Organisational characteristics (summary indicators) | |
		Enabling support environment type (ESE)	Community service provider (CSP)	ESE	CSP	Partnering typology	Participation in service delivery
1	Jharkhand	Centralised State Rural Water Supply Agency	Representative VWSC	55	67	Transactional	4. Interactive Participation
2	Madhya Pradesh	Hybrid (NGO)	Unregistered Society	95	67	Transactional	4. Interactive Participation
3	Odisha	External Agency	Registered Society	75	61	Operational	4. Interactive Participation
4	Chhattisgarh	Centralised State Rural Water Supply Agency	Representative VWSC	50	39	Transactional	3. Functional Participation
5	Meghalaya	Centralised State Rural Water Supply Agency	Representative VWSC	50	33	Transactional	4. Interactive Participation
6	Rajasthan	Centralised State Rural Water Supply Agency	Autonomous VWSC	50	33	Collaborative	3. Functional Participation
7	West Bengal	Hybrid (NGO)	Unregistered Society	30	33	Operational	4. Interactive Participation
8	Telangana	External Agency	Registered Society	90	75	Operational	4. Interactive Participation

Table A.4 continued

Case	State	Organisational typologies		Organisational characteristics (summary indicators)			
		Enabling support environment type (ESE)	Community service provider (CSP)	Professionalisation		Partnering typology	Participation in service delivery
				ESE	CSP		
9	Karnataka	Hybrid (Donor)	Representative VWSC	100	89	Transactional	4. Interactive Participation
10	Himachal Pradesh	Hybrid (Donor)	Unregistered Society	60	44	Collaborative	4. Interactive Participation
11	Punjab	Hybrid (Donor)	Registered Society	75	78	Collaborative	5. Self-mobilisation
12	Uttarakhand	External Agency	Registered Society	65	56	Operational	4. Interactive Participation
13	Kerala I – World Bank	Hybrid (Donor)	Registered Society	70	100	Transactional	5. Self-mobilisation
14	Kerala II – Local self-government	Decentralised Local Self-Government	Registered Society	100	64	Operational	4. Interactive Participation
15	Gujarat – WASMO Gandhinagar	Centralised State Rural Water Supply Agency	Registered Society	90	69	Operational	4. Interactive Participation
16	Gujarat – WASMO Kutch	Centralised State Rural Water Supply Agency	Registered Society	90	75	Operational	4. Interactive Participation

Table A.4 continued

Case	State	Organisational typologies		Professionalisation		Organisational characteristics (summary indicators)	
		Enabling support environment type (ESE)	Community service provider (CSP)	ESE	CSP	Partnering typology	Participation in service delivery
17	Tamil Nadu – Local self-government	Decentralised Local Self-Government	Representative VWSC	45	81	Collaborative	5. Self-mobilisation
18	Tamil Nadu – Public–private hybrid	Hybrid (Private)	Representative VWSC	75	69	Transactional	1. Passive Participation
19	Maharashtra	Centralised State Rural Water Supply Agency	Representative VWSC	55	33	Transactional	1. Passive Participation
20	Sikkim	Decentralised Local Self-Government	Representative VWSC	80	83	Operational	4. Interactive Participation

References

Abrams, L. J. (1998) Understanding Sustainability of Local Water Services. Retrieved March 2013 from www.thewaterpage.com/sustainability.htm.

Adank, M., Kumasi, T. C., Abbey, E., Dickinson, N., Dzansi, P., Atengdem, J. A., Chimbar, T. L. and Effah-Appiah, E. (2012) *The Status of Rural Water Supply Services in Ghana, A Synthesis of Findings from 3 Districts*. The Hague: IRC.

Adnan, S. (1992) *People's Participation, NGOs, and the Flood Action Plan: An Independent Review*. Dhaka, Bangladesh: Research and Advisory Services.

Amin. J., Denis, J., Harris, B., Ibenegbu, N., Javorszky, M., Maillot, E. and Tripp, S. (2015) *Determining Success in Community Managed Rural Water Supply Using Household Surveys*. Cranfield: Cranfield University Press.

Andrews, M., Pritchett, L. and Woolcock, M. (2013) Escaping Capability Traps Through Problem Driven Iterative Adaptation (PDIA). *World Development* 51: 234–244.

Arnstein, S. R. (1969) A Ladder of Citizen Participation. *Journal of the American Institute of Planners* 35(4): 216–224.

Asian Development Bank (2008) *Pakistan: Punjab Community Water Supply and Sanitation Sector Project*. Mandaluyong: Asian Development Bank.

Asthana, A. N. (2008) Decentralisation and Corruption: Evidence from Drinking Water Sector. *Public Administration and Development* 28(3): 181–189.

Bakalian, A. and Wakeman, W. (2009) *Post-Construction Support and Sustainability in Community-Managed Rural Water Supply: Case Studies in Peru, Bolivia and Ghana*. Washington, DC: World Bank–Netherlands Water Partnership (BNWP).

Banerjee, R. (2013) What Ails Panchayati Raj? *Economic and Political Weekly* 48(30): 172–176.

Baumann, E. (2006) Do Operation and Maintenance Pay? *Waterlines* 25(1): 10–12.

Bhatt, V. V. (1982) Development Problem, Strategy, and Technology Choice: Sarvodaya and Socialist Approaches in India. *Economic Development and Cultural Change* 31(1): 85–99.

Black, M. and Talbot, R. (2004) *Water: A Matter of Life and Health*. New Delhi: Oxford University Press.

Black, T.R. (2012) *Doing Quantitative Research in the Social Sciences: An Integrated Approach to Research Design, Measurement and Statistics*, 4th edition. London: Sage.

Bolt, E. and Fonseca, C. (2001) *Keep It Working: A Field Manual to Support Community Management of Rural Water Supply*. Technical Paper Series 36. Delft: IRC International Water and Sanitation Centre.

Boulenouar, J., Schweitzer, R. and Lockwood, H. (2013) *Mapping Sustainability Assessment Tools to Support Sustainable Water and Sanitation Service Delivery*. Wivenhoe: Aguaconsult.

Bovaird, T. (2007) Beyond Engagement and Participation: User and Community Coproduction of Public Services. *Public Administration Review* 67(5): 846–860.

Bovaird, T., Van Ryzin, G., Loeffler, E. and Parrado, S. (2015) Activating Citizens to Participate in Collective Co-Production of Public Services. *Journal of Social Policy* 44(1): 1–23.

Briscoe, J. and Malik, R. P. S. (2005) *India's Water Economy: Bracing for a Turbulent Future.* Washington, DC: World Bank.

Burns, T. and Stalker, G. M. (1961) *The Management of Innovation.* Oxford: Oxford University Press.

Burr, P. (2015) The Financial Costs of Delivering Rural Water and Sanitation Services in Lower-Income Countries. PhD thesis, Cranfield University, Cranfield.

Burr, P. and Fonseca, C. (2013) *Applying a Life-Cycle Costs Approach to Water: Costs and Service Levels in Rural and Small Town Areas in Andhra Pradesh (India), Burkina Faso, Ghana and Mozambique.* The Hague: IRC.

Census of India (2011a) *Census Data: Preliminary Results.* New Delhi: Census of India.

Census of India (2011b) *Main Source of Drinking Water 2001–2011.* New Delhi: Government of India.

Cepiku, D. and Giordano, F. (2014) Co-Production in Developing Countries: Insights from the Community Health Worker Experience. *Public Management Review* 16(3): 317–340.

Chambers, R. (1983) *Rural Development: Putting the Last First.* London: Routledge.

Chambers, R. (2008) *Revolutions in Development Inquiry.* London: Earthscan.

Chary Vedala, S., Jasthi, S. and Uddaraju, S. (2015a) *Users Becoming Managers of Water Supply; An Initiative of Water and Sanitation Management Organization, Gandhinagar District, Gujarat.* Community Water Plus Case Study Report 16. Hyderabad: Administrative Staff College of India.

Chary Vedala, S., Jasthi, S. and Uddaraju, S. (2015b) *Support to Community-Managed Rural Water Supply by the Water and Sanitation Management Organization in Kutch District, Gujarat.* Community Water Plus Case Study Report 17. Hyderabad: Administrative Staff College of India.

Chary Vedala, S., Jasthi, S. and Uddaraju, S. (2016a) *Professionally Managed – Community Owned Decentralized Drinking Water Service Delivery.* Community Water Plus Case Study 9. Hyderabad: Administrative Staff College of India.

Chary Vedala, S., Jasthi, S. and Uddaraju, S. (2016b) *Decentralisation Paving a Way for Efficient Service Delivery – A Case of Kodur Gram Panchayat, Kerala.* Community Water Plus Case Study 14. Hyderabad: Administrative Staff College of India.

Chary Vedala, S., Jasthi, S. and Uddaraju, S. (2016c) *Community Involvement in a Multi-Village Scheme in Amravati District, Maharashtra.* Community Water Plus Case Study Report 19. Hyderabad: Administrative Staff College of India.

Chowns, E. E. (2014) The Political Economy of Community Management: A Study of Factors Influencing Sustainability in Malawi's Rural Water Supply Sector. Thesis, 1 July. Retrieved from http://etheses.bham.ac.uk/5014/2/Decl_IS_Chowns.pdf.

Clasen, T. F. (2012) Millennium Development Goals Water Target Claim Exaggerates Achievement. *Tropical Medicine and International Health* 17(10): 1178–1180. doi:10.1111/j.1365-3156.2012.03052.x

Constitution of India (1950) *Constitution of India.* Delhi: Constitution of India.

Cullivan, D., Tippett, B., Edwards, D. B., Rosensweig, F. and McCaffery, J. (1988) *Guidelines for Institutional Assessment of Water and Wastewater Institutions.* WASH Technical Report No 37. Washington, DC: USAID.

Da Silva Wells, C., van Lieshout, R. and Uytewaal, E. (2013) Monitoring for Learning and Developing Capacities in the WASH Sector. *Water Policy* 15, 206–225.

Das, K. (2014) *The Sector Reforms Process in Rural Drinking Water and Sanitation: A Review of the Role of WASMO in Gujarat.* Gandhinagar, Gujarat: GIDR.

Datta, P. (2007) Devolution of Financial Power to Local Self-Governments: The 'Feasibility Frontier' in West Bengal. *South Asia Research* 27(1): 105–124.

Davis, J., Lukacs, H., Jeuland, M., Alvestegui, A., Soto, B., Lizárraga, G., Bakalian, A. and Wakeman, W. (2009) Sustaining the Benefits of Rural Water Supply Investments: Experience in Cochabamba and Chuquisaca, Bolivia. In A. Bakalian and W. Wakeman (eds), *Post-Construction Support and Sustainability in Community-Managed Rural Water Supply: Case Studies in Peru, Bolivia, and Ghana*, pp. 53–76. Washington, DC: World Bank–Netherlands Water Partnership.

Dayal, R., Van Wijk-Sijbesma, C. A. and Mukherjee, N. (2000) *Methodology for Participatory Assessments: With Communities, Institutions and Policy Makers: Linking Sustainability with Demand, Gender and Poverty*. Washington, DC: Water and Sanitation Program (WSP).

Deaton, A. (1997) *The Analysis of Household Surveys*. Baltimore, MD: Johns Hopkins University Press.

Demirjian, A. (2002) *Partnering in Support of International Development Initiatives: The INTO-SAI Case Study*. Ottawa: Consulting and Audit Canada.

Desai, M. (2006) *State Formation and Radical Democracy in India*. London: Routledge.

Deverill, P., Bibby, S., Wedgwood, A. and Smout, I. (2002) *Designing Water Supply and Sanitation Projects to Meet Demand in Rural and Peri-Urban Communities, Book 1: Concept, Principles and Practice*. Loughborough: WEDC.

Englebert, P. (2000) Pre-colonial Institutions, Post-colonial States, and Economic Development in Tropical Africa. *Political Research Quarterly* 53(1): 7–36.

Field, A. (2013) *Discovering Statistics Using SPSS*, 4th edition. London: Sage.

Fisher, M. B., Shields, K. F., Chan, T. U., Christenson, E., Cronk, R. D., Leker, H., Samani, D., Apoya, P., Lutz, A. and Bartram, J. (2015) Understanding Handpump Sustainability: Determinants of Rural Water Source Functionality in the Greater Afram Plains region of Ghana. *Water Resources Research* 51(10): 8431–8449.

Fonseca, C. (2014) Rural Water and Sanitation Household Costs in Lower-income Countries. PhD Thesis, Cranfield University, Cranfield.

Fonseca, C., Franceys, R. and Perry, C. (2010) *Guidelines for User Fees and Cost Recovery for Rural, Non-Networked, Water and Sanitation Delivery*. Tunis: African Development Bank.

Fonseca, C., Franceys, R., Batchelor, C., Mcintyre, P., Klutse, A., Komives, K. … Snehalatha, M. (2011) *Life-Cycle Costs Approach: Costing Sustainable Services*. The Hague: IRC.

Franceys, R. (2001) Promoting International Scientific and Technological Co-operation in Sustainable Water and Sanitation. Presented at OECD Conference on International Scientific and Technological Co-operation for Sustainable Development, Seoul.

Franceys, R. and Cavill, S. (2011) *A Comprehensive Literature Review on Charging for and Subsidising WASH Services*. New York: UNICEF.

Franceys, R. and Hutchings, P. (2017) Governance and Regulation of Water and Sanitation. In. Rieu-Clark et al. *Routledge Handbook of Water and Law*. Routledge: London.

Franceys, R., Cavill, S. and Trevett, A. (2016) Who Really Pays? A Critical Overview of the Practicalities of Funding Universal Access. *Waterlines* 35(1): 78–93.

Fuller, J. A., Goldstick, J., Bartram, J. and Eisenberg, J. N. S. (2016) Tracking Progress towards Global Drinking Water and Sanitation Targets: A Within and Among Country Analysis. *Science of the Total Environment* 541: 857–864.

Gerlach, E. and Franceys, R. (2009) 'Standpipes and Beyond' – a Universal Water Service Dynamic. *Journal of International Development* 22(4): 455–469.

Godfrey, S., Labhasetwar, P., Wate, S. and Pimpalkar, S. (2011) How Safe are the Global Water Coverage Figures? Case Study from Madhya Pradesh, India. *Environmental Monitoring and Assessment* 176(1–4): 561–574.

Government of Andhra Pradesh (2015) *Common Schedule of Rates as per A.P. Revised Standard Data for the Year 2015–16 (Effective from 1st June, 2015).* Hyderabad: Government of Andhra Pradesh.

Government of India (1860) *Societies Registration Act, 1860.* New Delhi: Government of India. Retrieved 14 April 2016 from www.mca.gov.in/Ministry/actsbills/pdf/Societies_Registration_Act_1860.pdf.

Government of India (1993) *The Constitution (Seventy-Third Amendment Act).* New Delhi: Government of India.

Government of India (2003) *Swajaldhara.* New Delhi: Department of Drinking Water Supply, Government of India.

Government of India (2009) *Report of Rajiv Gandhi National Drinking Water Mission.* New Delhi: Department of Drinking Water Supply, Government of India.

Government of India (2012a) *A Handbook for Gram Panchayats – To Help Them Plan, Implement, Operate, Maintain and Manage Drinking Water Security.* New Delhi: Ministry of Rural Development, Department of Drinking Water Supply and the Water and Sanitation Programme.

Government of India (2012b) *National Water Policy (Vol. 1).* New Delhi: Ministry of Water Resources.

Government of India (2013a) *Movement Towards Ensuring People's Drinking Water Security in Rural India: Framework for Implementation* (updated 2013). New Delhi: Ministry of Drinking Water and Sanitation/National Rural Drinking Water Programme/Rajiv Gandhi Drinking Water Mission.

Government of India (2013b) *National Water Policy 2012 RPRT.* New Delhi: Ministry of Water Resources.

Government of India (2015a) Government Constitutes National Institution for Transforming India (NITI). Retrieved 10 May 2015 from http://pib.nic.in/newsite/PrintRelease.aspx?relid=114268.

Government of India (2015b) Ranking of States by Devolution. Retrieved 4 May 2016 from www.panchayat.gov.in/hidden/-/asset_publisher/c6nFXzynlfcc/content/ranking-of-states-on-devolution-index/pop_up?_101_INSTANCE_c6nFXzynlfcc_viewMode=print.

Government of India (2015c) Ranking of States on Devolution Index. Retrieved 30 October 2015 from www.panchayat.gov.in/hidden/-/asset_publisher/c6nFXzynlfcc/content/ranking-of-states-on-devolution-index.

Gujarat State Disaster Management Authority (2003) *Gujarat Emergency Earthquake Reconstruction Project – Quarterly Progress Report,* July–September 2003. Government of Gujarat: Gandhinagar.

Harris, B., Brighu, U. and Poonia, R. (2016a) *Community Managed Water Supplies in Rural Jaipur: the Swajaldhara Scheme 15 Years On.* Community Water Plus Case Study 6. Jaipur: Malaviya National Institute of Technology.

Harris, B., Brighu, U. and Poonia, R. (2016b) *24x7 Water Supply in Punjab: International Funding for Local Action.* Community Water Plus Case Study 11. Jaipur: Malaviya National Institute of Technology.

Harris, B., Brighu, U. and Poonia, R. (2016c) *Community Water Classic: The Success of Community Managed Water Supplies in Himachal Pradesh with Limited On-going Support.* Community Water Plus Case Study Report 10. Jaipur: Malaviya National Institute of Technology.

Harris, D., Kooy, M. and Jones, L. (2011) *Analysing the Governance and Political Economy of Water and Sanitation Service Delivery.* London: Overseas Development Institute.

Harvey, P. A. and Reed, R. A. (2006) Community-Managed Water Supplies in Africa:

Sustainable or Dispensable? *Community Development Journal* 42(3): 365–378.

Heller, P., Harilal, K. N. and Chaudhuri, S. (2007) Building Local Democracy: Evaluating the Impact of Decentralization in Kerala, India. *World Development* 35(4): 626–648.

Hope, R. (2015) Is Community Water Management the Community's Choice? Implications for Water and Development Policy in Africa. *Water Policy* 17(4): 664.

Howard, G. and Bartram, J. (2013) *Domestic Water Quantity, Service Level and Health Executive summary*. Geneva, Switzerland: World Health Organization.

Hueso, A. and Bell, B. (2013) An Untold Story of Policy Failure: The Total Sanitation Campaign in India. *Water Policy* 15(6): 1001.

Hutchings, P. (2015) *Supporting Community Management in Morappur, Tamil Nadu*. Community Water Plus Case Study 18. Cranfield: Cranfield University Press.

Hutchings, P., Chan, M. Y., Cuadrado, L., Ezbakhe, F., Mesa, B., Tamekawa, C. and Franceys, R. (2015) A Systematic Review of Success Factors in the Community Management of Rural Water Supplies Over the Past 30 Years. *Water Policy* 17(5): 963.

Hutchings, P., Franceys, R., Mekala, S., Smits S. and James, A. J. (2016) Revisiting the History, Concepts and Typologies of Community Management for Rural Drinking Water Supply in India. *International Journal of Water Resources Development* 33(1): 152–169.

Hutton, G. and Varughese, M. (2016) *The Costs of Meeting the 2030 Sustainable Development Goal Targets on Drinking Water, Sanitation, and Hygiene*. Washington, DC: Water and Sanitation Program, World Bank.

Hutton, G., Haller, L. and Bartram, J. (2007) Global Cost–Benefit Analysis of Water Supply and Sanitation Interventions. *Journal of Water and Health* 5(4): 481–502.

James, A. J. (2004) *India's Sector Reform Projects and Swajaldhara Programme: A Case of Scaling up Community Managed Water Supply*. The Hague: IRC.

James, A. J. (2011) *Supporting Rural Water Supply: Assessing Progress Towards Sustainable Service Delivery (India)*. The Hague: IRC.

Javorszky, M., Dash, P. C. and Panda, P. K. (2015) *Understanding Resource Implications of the 'Plus' in Community Management of Rural Water Supply Systems in India: The Case of PHED*. Ranchi: Chhattisgarh.

Javorszky, M., Dash, P. C. and Panda, P. K. (2016) *Communal Action for Private Connections: Gram Vikas' Approach to Supporting Community Management of Rural Water Supplies in Odisha*. Community Water Plus Case Study 3. Ranchi: Xavier Institute for Social Service.

Johnson, C. (1982) *MITI and the Japanese Miracle: the Growth of Industrial Policy, 1925–1975*. Redwood, CA: Stanford University Press.

Johnson, C., Deshingkar, P. and Start, D. (2005) Grounding the State: Devolution and Development in India's Panchayats. *Journal of Development Studies* 41(6): 937–970.

Jones, S. (2011) Participation as Citizenship or Payment ? A Case Study of Rural Drinking Water Governance in Mali. *Water Alternatives* 4(1): 54–71.

Jones, S. (2013) How Can INGOs Help Promote Sustainable Rural Water Services? An Analysis of Wateraid's Approach to Supporting Local Governments in Mali. *Water Alternatives* 6(3): 350–366. Retrieved from www.scopus.com/inward/record.url?eid=2-s2.0-84886887823&partnerID=tZOtx3y1.

Jones, S. D. (2015) Bridging Political Economy Analysis and Critical Institutionalism: An Approach to Help Analyse Institutional Change for Rural Water Services. *International Journal of the Commons* 9(1): 65–86. Retrieved 27 July 2015 from www.thecommonsjournal.org/index.php/ijc/article/view/URN%3ANBN%3ANL%3AUI%3A10-1-116921/472.

Joshi, A. and Moore, M. (2004) Institutionalised Co-production: Unorthodox Public Service Delivery in Challenging Environments. *Journal of Development Studies* 40(4): 31–49.

Joshi, D. (2003) *Secure Water? Poverty, Livelihoods and Demand-Responsive Approaches*. London: Overseas Development Institute. Retrieved 21 April 2016 from www.odi.org/sites/odi.org.uk/files/odi-assets/publications-opinion-files/1618.pdf.

Kelsall, T. (2011) Rethinking the Relationship between Neo-patrimonialism and Economic Development in Africa. *IDS Bulletin* 42(2): 76–87.

Kerala State Planning Board (2009) *An Evaluation Study on Jalanidhi Projects in Kerala*. Thiruvananthapuram, Kerala: Kerala State Planning Board.

Kleemeier, E. (2000) The Impact of Participation on Sustainability: An Analysis of the Malawi Rural Piped Scheme Program. *World Development* 28(5): 929–944.

Koehler, J., Thomson, P. and Hope, R. (2015) Pump-Priming Payments for Sustainable Water Services in Rural Africa. *World Development* 74: 397–411.

Kohli, A. (2012) *Poverty Amid Plenty in the New India*. New York: Cambridge University Press.

Lammerink, M. and de Jong, D. (1999) *Community Water Management*. PLA Notes 35. London: International Institute for Environment and Development.

Lieberson, S. and O'Connor, J. F. (1972) Leadership and Organizational Performance: A Study of Large Corporations. *American Sociological Review* 37(2): 117.

Lockwood, H. (2002) *Institutional Support Mechanisms for Community-Managed Rural Water Supply and Sanitation Systems in Latin America*. Environmental Health Project (EHP) Strategic Report 6. Washington, DC: USAID. Retrieved 5 May 2015 from http://pdf.usaid.gov/pdf_docs/PNACR786.pdf.

Lockwood, H. (2004) *Scaling Up the Community Management of Rural Water Supply*. The Hague: IRC.

Lockwood, H. and Smits, S. (2011) *Supporting Rural Water Supply: Moving Towards a Service Delivery Approach*. Rugby: Practical Action Publishing.

Lockwood H., Bakalian, A. and Wakeman, W. (2003) *Assessing Sustainability in Rural Water Supply: The Role of Follow-Up Support to Communities; Literature Review and Desk Review of Rural Water Supply and Sanitation Project Documents*. Washington, DC: World Bank.

Madrigal-Ballestero, R., Alpízar, F. and Schlüter, A. (2013) Public Perceptions of the Performance of Community-Based Drinking Water Organizations in Costa Rica. *Water Resources and Rural Development* 1–2: 43–56.

Mandara, C. G., Butijn, C. and Niehof, A. (2013) Community Management and Sustainability of Rural Water Facilities in Tanzania. *Water Policy* 15(S2): 79.

Marks, S. J. and Davis, J. (2012) Does User Participation Lead to Sense of Ownership for Rural Water Systems? Evidence from Kenya. *World Development* 40(8): 1569–1576.

McCommon, C., Warner, D. and Yohalem, D. (1990) *Community Management of Rural Water Supply and Sanitation Services*. WASH Technical Report 67. Washington, DC: USAID.

McIntyre, P., Casella, D., Fonseca, C. and Burr, P. (2014) *Supporting Water Sanitation and Hygiene Services for Life: Uncovering the Real Costs of Water and Sanitation*. The Hague: IRC. Retrieved 22 April 2016 from www.ircwash.org/sites/default/files/2014_priceless_mcintyreetal_1.pdf.

Ministry of Drinking Water and Sanitation (2016) Differences in IMIS and Census 2011: Village List. Retrieved 18 July 2016 from http://indiawater.gov.in/imisreports/Reports/Alert/rpt_RWS_diffNoWithCensus2011Vill_S.aspx?Rep=0&RP=YCoverage.

Ministry of Drinking Water Supply (MDWS) (2013) Overview of National Drinking Water Security Pilot Programme. Retrieved 22 August 2013 from www.mdws.gov.in/sites/default/files/NDWSP%281%29.pdf.

Moriarty, P., Smits, S., Butterworth, J. and Franceys, R. (2013) Trends in Rural Water Supply: Towards a Service Delivery Approach. *Water Alternatives* 6(3): 329–349.

MOUD (undated) *Handbook on Service Level Benchmarking.* New Delhi: Ministry of Urban Development,.

Nayar, V. (2006) *Democratisation of Water Management: Establishing a Paradigm Shift in Water Sector.* Tamil Nadu: TWAD Board.

Nayar, V. and James, A. J. (2010) Policy Insights on User Charges from a Rural Water Supply Project: A Counter-intuitive View from South India. *International Journal of Water Resources Development* 26(3): 403–421.

Nelson, F. and Agrawal, A. (2008) Patronage or Participation? Community-Based Natural Resource Management Reform in Sub-Saharan Africa. *Development and Change* 39(4): 557–585.

Norman, G., Fonseca, C. and Trémolet, S. (2015) *Domestic Public Finance for WASH: What, Why, How?* Retrieved 4 August 2016 from www.publicfinanceforwash.com/sites/default/files/uploads/Finance_Brief_1_-_Domestic_public_finance_for_WASH.pdf.

North, D. C. (1990) Institutions and Economic Growth: An Historical Introduction. *World Development* 17(9): 1319–1332.

OECD (2009) *Managing Water for All: An OECD Perspective on Pricing and Financing.* Paris: OECD. Retrieved 22 April 2016 from www.oecd.org/tad/sustainable-agriculture/44476961.pdf.

Ofgem (2016) Electricity. Retrieved 24 February 2016 from www.ofgem.gov.uk/electricity.

Onda, K., LoBuglio, J. and Bartram, J. (2012) Global Access to Safe Water: Accounting for Water Quality and the Resulting Impact on MDG Progress. *International Journal of Environmental Research and Public Health* 9(3): 880–894.

Padawangi, R. (2010) Community-Driven Development as a Driver of Change: Water Supply and Sanitation Projects in Rural Punjab, Pakistan. *Water Policy* 12(S1): 104–120.

Parel, A. (2011) Gandhi and the State. In J. M. Brown and A. Parel (eds), *The Cambridge Companion to Gandhi*, pp. 154–172. Cambridge: Cambridge University Press.

Parker, A. H., Youlten, R., Dillon, M., Nussbaumer, T., Carter, R. C., Tyrrel, S. F. and Webster, J. (2010) An Assessment of Microbiological Water Quality of Six Water Source Categories in North-east Uganda. *Journal of Water and Health* 8(3): 550–560.

Paul, S. (1987) Community Participation in Development Projects: The World Bank Experience. In *Readings in Community Participation*. Washington, DC: EDI.

Peters, D. H., Paina, L. and Schleimann, F. (2013) Sector-Wide Approaches (SWAps) in Health: What Have We Learned? *Health Policy Plan* 28(8): 884–890.

Peters, D. and Chao, S. (1998) The Sector-Wide Approach in Health: What Is It? Where Is It Leading? *International Journal of Health Planning and Management* 13(2): 177–190.

Petticrew, M. and Roberts, H. (eds) (2006) *Systematic Reviews in the Social Sciences.* Oxford: Blackwell Publishing.

Planning Commission (2014) *Gross State Domestic Product (GSDP) at Current Prices (as on 15-03-2012).* New Delhi: Planning Commission, Government of India.

Postma, L., James, A. J. and van Wijk, C. (2004) QIS: A New Participatory Management Tool to Assess and Act on Field Reality. Presented at People-Centred Approaches to Water and Environmental Sanitation, 30th WEDC International Conference, Loughborough, UK.

Pretty, J. (1994) Alternative Systems of Inquiry for a Sustainable Agriculture. *IDS Bulletin* 25: 37–49. University of Sussex: IDS.

Pritchett, L., Woolcock, M. and Andrews, M. (2013) Looking Like a State: Techniques of Persistent Failure in State Capability for Implementation. *Journal of Development Studies* 49(1): 1–18.

Prokopy, L. S. (2009) Determinants and Benefits of Household Level Participation in Rural Drinking Water Projects in India. *Journal of Development Studies* 45(4): 471–495.

Rajesh, K. and Thomas, M. B. (2012) Decentralization and Interventions in the Health Sector. *Journal of Health Management* 14(4): 417–433.

Ramamohan Roa, M. S. and Raviprakash, M. S. (2016a) *Empowered Community: Secured Safe Water Supply In Parts of Dhar District, Madhya Pradesh.* Chennai: Centre for Excellence in Change.

Ramamohan Roa, M. S. and Raviprakash, M. S. (2016b) *Jal Nirmal and Beyond: Supporting the Community Management of Rural Water Supply in Belagavi District, Karnataka.* Chennai: Centre for Excellence in Change.

Ratna Reddy, V. and Batchelor, C. (2012) Cost of Providing Sustainable Water, Sanitation and Hygiene (WASH) Services: An Initial Assessment of a Life-Cycle Cost Approach (LCCA) in Rural Andhra Pradesh, India. *Water Policy* 14(3): 409–429.

Ratna Reddy, V., Ramamohan Roa, M. S. and Venkataswamy, M. (2010) *'Slippage': The Bane of Rural Drinking Water Sector.* The Hague: IRC.

Reserve Bank of India (2015) *Handbook of Statistics on Indian Economy.* New Delhi: Reserve Bank of India.

Robinson, L. (2003) Consultation: What Works? Presentation to Local Government Public Relations Conference, Wollongong, Australia.

Robinson, L. and Nolan-ITU (2002) *Pro-active Public Participation for Waste Management in Western Australia. Part 1 – Strategic Rationale: Why Should Communities Participate in Waste Management Technology and Siting Decisions?* West Leederville: Western Australian Local Government Association (WALGA) and Waste Education Strategy Integration Group (WESIG).

Robinson, M. (2007) Introduction: Decentralising Service Delivery? Evidence and Policy Implications. *IDS Bulletin* 38(1): 1–6.

Rout, S. (2014) Institutional Variations in Practice of Demand Responsive Approach: Evidence from Rural Water Supply in India. *Water Policy* 16(4): 650.

Ruparelia, S., Reddy, S., Harriss, J. and Corbridge, S. (2011) *Understanding India's New Political Economy.* Abingdon: Routledge.

RWSN (2009) *Myths of the Rural Water Supply Sector.* St Gallen, Switzerland: Rural Water Supply Network.

Saraswathy, R. (2015) *Kathirampatti Village Panchayat, Tamil Nadu Rural Water Supply.* Community Water Plus Case Study Report 17. Chennai: Centre for Excellence in Change.

Saraswathy, R. (2016a) *Nenmeni Sudha Jala Vitharana Society (NSJVS) Kerala: Professionalised Management of Water Supply by Community.* Community Water Plus Case Study Report 13. Chennai: Centre for Excellence in Change.

Saraswathy, R. (2016b) *The Dorbars and Gravity-Fed Piped Water Supply in Meghalaya.* Community Water Plus Case Study 5. Chennai: Centre for Excellence in Change.

Saraswathy, R. (2016c) *Community-Managed Gravity-Fed Piped Water Supply in Himalayan Sikkim.* Community Water Plus Case Study Report 20. Chennai: Centre for Excellence in Change.

Schouten, T. and Moriarty, P. (2003) *Community Water, Community Management: From System to Service in Rural Areas.* London: ITDG Publishing.

Schweitzer, R. W. and Mihelcic, J. R. (2012) Assessing Sustainability of Community Management of Rural Water Systems in the Developing World. *Journal of Water, Sanitation and Hygiene for Development* 2(1): 20–30.

Sen, S. (1999) Some Aspects of State-NGO Relationships in India in the Post-Independence Era. *Development and Change* 30(2): 327–355.

Sharma, V. (2015) Are BIMARU States Still Bimaru? *Economic and Political Weekly* 50(18): 58–63.

Smits, S. and Mekala, S. (2015) *The Effects and Costs of Support to Community-Managed Handpumps in Patharpratima, West Bengal.* Community Water Plus Case Study 6. The Hague: IRC.

Smits, S., Verhoeven, J., Moriarty, P., Fonseca, C. and Lockwood, H. (2011) *Arrangements and Costs of Support to Rural Water Service Providers.* The Hague: IRC.

Smits, S., Rojas, J. and Tamayo, P. (2013) The Impact of Support to Community-Based Rural Water Service Providers: Evidence from Colombia. *Water Alternatives* 6(3): 384–404.

Smits, S., Franceys, R., Mekala, S. and Hutchings, P. (2015) *Understanding the Resource Implications of the 'Plus' in Community Management of Rural Water Supply Systems in India: Concepts and Research Methodology.* The Hague: IRC.

Smits, S., Shiva, R. and Kapur, D. (2016) *Support to Community-Managed Rural Water Supplies in the Uttarakhand Himalayas – the Himmotthan Water Supply and Sanitation Initiative.* The Hague: IRC.

Snehalatha M., Busenna P., Ratna Reddy, V. and Anitha, V. (2011) *Rural Drinking Water Service Levels: A Study of Andhra Pradesh, South India.* CESS Working Paper No.13 December, 2011. Hyderabad: Centre for Economic and Social Studies.

Srivastava, S. (2012) Swajaldhara: 'Reversed' Realities in Rural Water Supply in India. *IDS Bulletin* 43(2): 37–43.

Stalker Prokopy, L. and Thorsten, R. (2009) Post-Construction Support and Sustainability in Rural Drinking Water Projects in Cuzco, Peru. In A. Bakalian and W. Wakeman (eds), *Post-Construction Support and Sustainability in Community-Managed Rural Water Supply: Case Studies in Peru, Bolivia, and Ghana*, pp. 17–51. Washington, DC: World Bank–Netherlands Water Partnership.

Sutton, S. (2005) The Sub-Saharan Potential for Household Level Water Supply Improvement. Presented at 31st WEDC International Conference, Kampala, Uganda.

Sutton, S. (2008) *An Introduction to Self-Supply: Putting the User First.* St Gallen, Switzerland: Rural Water Supply Network.

Tamayo, S. P. and García, M. (2006) Estrategia estatal para el fortalecimiento de entes prestadores de servicios públicos en el pequeño municipio y la zona rural: El programa cultura empresarial adelantado en Colombia. In F. Quiroz, N. Faysse and R. Ampuero (eds), *Apoyo a la gestión de Comités de Agua Potable; experiencias de fortalecimiento a comités de agua potable con gestión comunitaria en Bolivia y Colombia.* Cochabamba, Bolivia: Centro Agua–UMSS.

Thomas, S. K. (2013) Gram Vikas: An Overview. Presented at Community Water Plus Stakeholder Meeting, New Delhi, India.

Transparency International (2014) Corruption Perceptions Index 2014. Transparency International Secretariat, Berlin. Retrieved 15 January 2015 from www.transparency.org/cpi2014

UN Water and WHO (2014) Investing in Water and Sanitation: Increasing Access, Reducing Inequalities. Retrieved 22 April 2016 from www.who.int/water_sanitation_health/glaas/en.

Union Ministry of Water Resources (2016) *Draft National Water Framework Bill.* Draft of 16 May 2016. New Delhi: Government of India.

United Nations (2015) Sustainable Development Goals. New York: United Nations. Retrieved 8 December 2015 from https://sustainabledevelopment.un.org/?menu=1300.

Van den Broek, M. and Brown, J. (2015) Geoforum Blueprint for Breakdown? Community Based Management of Rural Groundwater in Uganda. *Geoforum* 67: 51–63.

Vanderwal, J. H. (1999) Negotiating Restoration: Integrating Knowledges on the Alouette River, British Columbia. MSc thesis, University of British Columbia.

WAMS (2016) Water Asset Management System Andhra Pradesh. Rural Water Supply and Sanitation, Government of Andhra Pradesh: Amaravati.

WaterAid (2008) *Think Local, Act Local: Effective Financing of Local Governments to Provide Water and Sanitation Services.* London: WaterAid.

Weber, M. (1978) *Economy and Society: An Outline of Interpretive Sociology.* Berkeley, CA: University of California Press.

Westcoat, J., Fletcher, S. and Novellino, M. (2016) National Rural Drinking Water Monitoring: Progress and Challenges with India's IMIS Database. *Water Policy* 18(4).

Whittington, D., Briscoe, J., Mu, X. and Barron, W. (1990) Estimating the Willingness to Pay for Water Services in Developing Countries: A Case Study of the Use of Contingent Valuation Surveys in Southern Haiti. *Economic Development and Cultural Change* 38(2): 293–311.

Whittington, D., Davis, J., Prokopy, L., Komives, K., Thorsten, R., Lukacs, H., Bakalian, A. and Wakeman, W. (2009) How Well is the Demand-Driven, Community Management Model for Rural Water Supply Systems Doing? Evidence from Bolivia, Peru and Ghana. *Water Policy* 11(6): 696.

WHO and UNICEF (2012) *Progress on Drinking Water and Sanitation: 2012 Update.* New York: WHO and UNICEF.

WHO and UNICEF (2013) *Progress on Drinking Water and Sanitation: 2013 Update.* New York: WHO and UNICEF.

WHO and UNICEF (2015) *Progress on Drinking Water and Sanitation: 2015 Update.* New York: WHO and UNICEF.

WHO and UNICEF (2016) *Progress on Drinking Water and Sanitation: 2016 Update.* New York: WHO and UNICEF.

Wilcox, D. (1994) *The Guide to Effective Participation.* London: Partnership Books.

World Bank (2001a) *La asociacion de usuarios en la gestion de servicios de agua en localidades rurales multiples: el caso de el Ingenio en Ica, Nasca, Perú.* Washington, DC: World Bank Water and Sanitation Program – Andean Region.

World Bank (2001b) *Multi-Village Water Supply Schemes in India.* Discussion paper. Washington, DC: World Bank.

World Bank (2002) *Willingness to Charge and Willingness to Pay: The World Bank-Assisted China Rural Water Supply and Sanitation Program.* Washington, DC: World Bank.

World Bank (2004) *World Development Report 2004: Making Services Work for Poor People.* Washington, DC: World Bank.

World Bank (2016) GDP per Capita, PPP (Current International $). Retrieved 27 April 2016 from http://data.worldbank.org/indicator/NY.GDP.PCAP.PP.CD.

WSP (2002) *Sustainability of Rural Water Supply Projects: Lessons from the Past – South Asia Region.* Washington, DC: Water and Sanitation Program, World Bank.

Yin, R. K. (2003) *Case Study Research Design and Methods,* 3rd edition. London: Sage.

Index

3Ts 19

accessibility 160
accountability 149, 150, 151
Andhra Pradesh: Rural Water Supply
(RWS) 89, 181–2; state-level monitoring
181–3; water kiosks 91
appraisals 75
appropriate technology 11
autonomous VWSC 150, 151, 155; capital
expenditure 153; financial costs 170;
overall costs 154
autonomy 111, 144, 145

Block Resource Centres 61–2, 75
bulk water management models 109, 110

capital expenditure (CapEx) 20, 57;
autonomous VWSCs 153; cost sharing
168; developmental states 107; enabling
support environment categories 146;
financial costs by organisation type
169–70; financial incentives 113;
hardware 166, 167; high level of 110;
high performance groups 172, 173; low
performance groups 172, 173; medium
performance groups 173; mountainous
regions 120; neo-patrimonial states 70,
72; passive participation 110; registered
societies 153; representative VWSCs 153,
171; social-democratic states 92;
software 166, 167; unregistered societies
153; WASHCost project 167
capital investment phase 8; bureaucratic
partnering 218; collaborative partnering
217; consultative partnering 217;
contributory partnering 217; functional
participation 215; interactive
participation 215; operational partnering

217; participation by consultation 216;
passive participation 216; self-
mobilisation 215; transactional
partnering 218
capitalist production: interventionist
approach 44; laissez-faire approach 44
capital maintenance expenditure
(CapManEx) 20, 57; hardware 166, 167;
neo-patrimonial states 71; software 166,
167
capital maintenance phase 8, 215–16;
bureaucratic partnering 218;
collaborative partnering 217;
consultative partnering 217;
contributory partnering 217; functional
participation 215; interactive
participation 215; operational partnering
217; participation by consultation 216;
passive participation 216; self-
mobilisation 215; transactional
partnering 218
centralised government support 134–6;
institutional assessment 144, 145;
organisational arrangements 154;
participation 142; partnering assessment
143
Chhattisgarh, case study 52, 62;
government subsidies 66; household
service level 69; limited autonomy of
VWSC 72; organisational characteristics
73; recurrent costs 70; responsibility of
PHED 66; service levels and financial
costs 158; women engineers 197;
women's representation in village
committee membership 193
community contribution: capital
expenditure 70, 72, 167–8, 170, 171,
172, 173; demand-responsive approach
71–2, 170; IQR 168; link with service-

For Product Safety Concerns and Information please contact our EU
representative GPSR@taylorandfrancis.com
Taylor & Francis Verlag GmbH, Kaufingerstraße 24, 80331 München, Germany

www.ingramcontent.com/pod-product-compliance
Ingram Content Group UK Ltd.
Pitfield, Milton Keynes, MK11 3LW, UK
UKHW021003180425
457613UK00019B/799